AF848629

Asian Mobilities Consumption in a Changing Arctic

This book provides an in-depth examination of the growing Asian tourism market and consumption in Arctic destinations.

Through five parts, the book covers Asian mobilities consumption as an extension of Arctic international politics, the transportation sector and green cruise tourism, and ethnicity, culture, and history. It contributes to further understanding of the impacts of increased tourism in these polar regions by exploring climate change, debates around emerging economies, and global power roles in the political, socio-economic, security, and legal issues of the Arctic and Antarctic and associated polar strategies and policy. By drawing on a range of disciplines and with contributions from experts in Arctic destinations or who are associated with the Arctic, it further provides a holistic framing of emerging demand and mobility patterns of Asian tourists in a polar context.

Asian Mobilities Consumption in a Changing Arctic will be valuable reading for students and academics across the fields of tourism, economics, sustainability, and development studies as well as other social science disciplines.

Young-Sook Lee is professor in tourism management at the School of Business and Economics, UiT The Arctic University of Norway. As a Korean-Australian (moved to Australia after her first degree), she completed her PhD in Sociology (applied in tourism) at the University of Queensland, Australia. She worked as a tourism management academic in Australia (Griffith University) for over 15 years before relocating to Norway in 2014.

Routledge Advances in Tourism
Edited by **Stephen Page**, *Hertfordshire Business School, University of Hertfordshire, UK*

Tourism, Resilience and Sustainability
Adapting to Social, Political and Economic Change
Edited by Joseph M. Cheer and Alan A. Lew

Slow Tourism, Food and Cities
Pace and the Search for the 'Good Life'
Edited by Michael Clancy

Lifelong Learning for Tourism
Concepts, Policy and Implementation
Edited by Violet V. Cuffy, David Airey and Georgios C. Papageorgiou

Hosting the Olympic Games
Uncertainty, Debates, and Controversy
Edited by Marie Delaplace and Pierre-Olaf Schut

Encountering Nazi Tourism Sites
Derek Dalton

Evaluating the Local Impacts of the Rio Olympics
Edited by Marcelo Neri

Mountain Resort Marketing and Management
Armelle Solelhac

Ageing and the Visitor Economy
Global Challenges and Opportunities
Stephen J. Page and Joanne Connell

Asian Mobilities Consumption in a Changing Arctic
Edited by Young-Sook Lee

For more information about this series, please visit: www.routledge.com/advances-in-tourism/book-series/SE0538

Asian Mobilities Consumption in a Changing Arctic

Edited by Young-Sook Lee

Routledge
Taylor & Francis Group
LONDON AND NEW YORK

First published 2022
by Routledge
2 Park Square, Milton Park, Abingdon, Oxon OX14 4RN

and by Routledge
605 Third Avenue, New York, NY 10158

Routledge is an imprint of the Taylor & Francis Group, an informa business

© 2022 selection and editorial matter, Young-Sook Lee; individual chapters, the contributors

The right of Young-Sook Lee to be identified as the author of the editorial material, and of the authors for their individual chapters, has been asserted in accordance with sections 77 and 78 of the Copyright, Designs and Patents Act 1988.

All rights reserved. No part of this book may be reprinted or reproduced or utilised in any form or by any electronic, mechanical, or other means, now known or hereafter invented, including photocopying and recording, or in any information storage or retrieval system, without permission in writing from the publishers.

Trademark notice: Product or corporate names may be trademarks or registered trademarks, and are used only for identification and explanation without intent to infringe.

British Library Cataloguing-in-Publication Data
A catalogue record for this book is available from the British Library

Library of Congress Cataloging-in-Publication Data
Names: Lee, Young-Sook, editor.
Title: Asian mobilities consumption in a changing Arctic/edited by Young-Sook Lee.
Description: Abingdon, Oxon; New York, NY: Routledge, 2022. |
Series: Routledge advances in tourism | Includes bibliographical references and index.
Identifiers: LCCN 2021019263 (print) | LCCN 2021019264 (ebook)
Subjects: LCSH: Tourism–Polar regions. | Asians–Travel–Polar regions. |
Consumption (Economics)–Asia. | Polar regions–Economic conditions.
Classification: LCC G155.P65 A75 2022 (print) | LCC G155.P65 (ebook) |
DDC 338.4/791089950113–dc23
LC record available at https://lccn.loc.gov/2021019263
LC ebook record available at https://lccn.loc.gov/2021019264

ISBN: 978-0-367-48361-6 (hbk)
ISBN: 978-1-032-10603-8 (pbk)
ISBN: 978-1-003-03951-8 (ebk)

DOI: 10.4324/9781003039518

Typeset in Times New Roman
by Deanta Global Publishing Services, Chennai, India

To my Mother and Father from Young-Sook

Contents

Figures

Tables

Contributors

Emma Björner (University of Gothenburg and Lund University, Sweden), PhD, is a researcher at Gothenburg Research Institute and senior lecturer in the Department of Strategic Communication. Her research and teaching are in the areas of sustainable tourism, sustainable cities, place branding, marketing and branding, with a focus on Europe and China. She has published in journals such as *Cities*, *Place Branding*, and *Public Diplomacy*, and the *Journal of Place Management and Development*.

Rosemary Black holds a PhD from Monash University and is Associate Professor in Interpretation and Ecotourism at Charles Sturt University. She teaches and undertakes research in sustainable tourism, tour guiding, environmental communication, and social dimensions of natural resource management. Prior to becoming an academic, she worked in protected area management, adventure travel, and community conservation. She has published six books and more than 60 refereed publications. Rosemary undertakes applied research with industry partners, including protected area management, tourism agencies, and community-based organisations. Rosemary is on the editorial board of three international tourism journals.

David Chapman is a British architect and town planner. He made his doctorate in Architecture, researching the urban design of winter cities and the relationship between human movement, urban design, and climate. Today, David is Associate Senior Lecturer at Lulea University of Technology (Sweden), where he teaches courses in Architecture, Urban Design, and Town Planning.

Wilson (Wai-Yin) Cheung is a PoSSUM scientist-astronaut candidate, a fellow of the Royal Geographical Society (RGS), and a member of The Explorers Club and Scientific Committee on Antarctic Research, Humanities & Social Sciences Expert Group. Since 2011, he has worked over 1,000 field days as a polar expedition and mountain and ski safety consultant for international scientists and visitors in extremely remote areas to ensure a secure wilderness experience for international visitors. The polar region is part of his life. He has lived in East Greenland with local hunters and learnt their way of living. In each Antarctic season, he spends over five months on the Antarctic continent and peninsula region.

Matteo Chiesa is a material scientist working on creating and implementing technologies necessary to adapt the current energy system into a competitive and secure one and designs and investigates engineered nanomaterials for solar energy conversion and storage systems. Currently, the focus of Matteo's research is on tackling specific challenges related to the deployment of renewable energy under Arctic conditions through a synthetic and symbiotic approach. Matteo cultivates interdisciplinarity and combines technologies to create optimal solutions while ensuring their harmonious integration in a societal context to achieve sustainable living and wellbeing of people.

Elena Dybtsyna (Nord University, Norway), PhD, is Associate Professor in Business Administration in the Business School and High North Centre for Business and Governance. She is teaching and doing research in areas connected to international business and governance in the Arctic/High North, management control in public and private sectors, the circular economy, and smart cities. She has recently co-authored a chapter in the Palgrave Handbook of Arctic Policy and Politics.

Odin Foldvik Eikeland is a PhD student in renewable energy with expertise in solar energy and energy analytics, working to analyse current energy-related challenges, and develop potential solutions that are necessary for the transition to a competitive, secure, and more sustainable energy system. Odin's main research focus in the PhD project is to analyse and identify the variables that cause faults in the power grid by use of advanced machine learning methodologies and then to develop new solutions that are innovative, secure, and able to provide a stable power supply for local communities.

Firouz Gaini is Professor in anthropology at the University of the Faroe Islands. His publications cover research fields such as youth cultures, changing identities, local knowledge, small island communities, and sustainable tourism in the North Atlantic. He is co-author of the book Gender and Island Communities (Routledge 2020). He has conducted fieldwork in the Faroe Islands, Greenland, France, and Japan. He is the research leader in the Faculty of History and Social Sciences, at the University of the Faroe Islands.

Guðrún Þóra Gunnarsdóttir is Director of the Icelandic Tourism Research Centre. She holds an MBA degree in tourism management from the University of Guelph, Ontario, Canada and a MA degree in comparative literature from the University of Oregon, USA. Gudrun has more than two decades of experience in teaching and researching tourism. She led the development of the Department of Rural Tourism at Holar University that was established in 1996. Her research interests have focused on tourism in rural areas and destination development.

Gestur Hovgaard is Professor and Head of Department at Ilisimatusarfik, University of Greenland and is also affiliated with the University of Faroe

Islands. He holds a Master's degree in public administration and a PhD in social science from Roskilde University, Denmark. His publications cover local and regional development and planning, innovation, and public-sector organisation, mainly in a Nordic Atlantic context. His recent research focuses on public planning, educational planning, tourism planning, and economic and social history.

Matouš Jelínek is a methodologist of qualitative research and analysis at the Research Institute of Labour and Social Affairs—RILSA and a PhD candidate in the Department of Sociology of Masaryk University. His current research is concerned with methods and practices of ethnic identification regarding ethnic data collection and the intersections of ethnicity, gender, and class in the context of social inequality and marginalisation. He has carried out several ethnographical research projects in Roma communities and other research projects related to social services and for-profit care agencies.

Tuomas Kiiski received an MSc and DSc in Operations & Supply Chain Management from the Turku School of Economics at the University of Turku, Finland. His research interests are in quantitative modelling and maritime economics. He has published several articles in top journals in the field.

Mariia Kobzeva has a PhD in Political Science and is a postdoctoral fellow in global Arctic studies in the Department of Social Sciences, UiT Norwegian Arctic University. Mariia is focused on Sino-Russian relations in the Arctic amid new balances of power as well as on Russian and Chinese foreign policies.

Eva Kotašková is a PhD candidate in the Department of Sociology of Masaryk University. Her current research is an ethnography of interacting with the environment within the tourist industry in the Svalbard archipelago, focusing mainly on the anthropology of nature-culture relations. She is a founding member of the Svalbard Social Science Initiative. She has also been educated and works as a nature guide in Svalbard.

Agneta Larsson is Senior Lecturer at the Luleå University of Technology (Sweden). Her practice and research focus on risk, safety, and health. Specifically, how interactions between environmental conditions, human capacity, and behaviour affect how a person feels comfortable moving in the outdoor environment. She teaches courses in Public Health, Integrative hysiology, and Physiotherapy.

Erik Lundberg (University of Gothenburg, Sweden), PhD, is a researcher and Lecturer in Marketing and at the Centre for Tourism, University of Gothenburg, Sweden. In 2014, he received his PhD, in which he describes and analyses tourism and event impacts from a sustainable development perspective. He has published in journals such as *Tourism Management*, *Journal of Sustainable Tourism*, *Scandinavian Journal of Hospitality and Tourism Management*, and *Event Management*.

Line Mathisen has a PhD in tourism from UiT The Arctic University of Norway. Line Mathisen is Associate Professor in the School of Business and Economics, UiT The Arctic University of Norway. Her current research interests are connected to collaboration logics and value creation in development processes, which include areas such as tourism, regional development, entrepreneurship, and innovation.

Minna-Liina Ojala (M.Sc. in Geography) is Project Researcher in the Turku School of Economics (Operations and Supply Chain Management) at the University of Turku, Finland. She is finalising her doctoral thesis on knowledge creation processes among multicultural and mobile groups with case studies reaching from the Arctic to Sub-Saharan Africa.

Juulia Räikkönen (D.Sc. Econ. & Bus. Admn.) is Adjunct Professor and Post-Doctoral Research Fellow at the University of Turku. She works in the Turku School of Economics (Economic Geography) as well as the Biodiversity Unit. Her research interests cover a wide variety of topics related to tourism, leisure consumption, and experience management.

Nathan Reigner, PhD, is an applied social scientist and Research Assistant Professor in the Recreation, Park, and Tourism Management Department at the Pennsylvania State University. Nathan's work helps protected areas and tourism destinations design and develop sustainable visitor and tourism management systems. He has worked extensively within the US national park and forest systems and in the Nordic Arctic, former Soviet, and middle eastern countries. Nathan holds a PhD in natural resource management from the University of Vermont and was an Arctic Fulbright Scholar in Iceland in 2019.

Zdenka Sokolíčková is a guest researcher at the Department of Social Anthropology, University of Oslo, mentored by Thomas Hylland Eriksen during her research project boREALIFE: Overheating in the High Arctic (fieldwork in Longyearbyen 2019–2021). She is interested in the interconnectedness of environmental, economic, and societal change in the globalised and overheated world. She is Board Leader of the Svalbard Social Science Initiative and Associate Professor at the Department of Studies in Culture and Religion, University of Hradec Králové, Czech Republic. In March 2021, she joined the interdisciplinary team at the Arctic Centre in Groningen, the Netherlands, as a postdoc in the project SVALUR: Understanding Resilience and Long-Term Environmental Change in the High Arctic: Narrative-Based Analyses from Svalbard.

Siri Ulfsdatter Søreng is Associate Professor at the School of Business and Economics at UiT the Arctic University of Norway. She has a PhD in natural resource management from the Norwegian College of Fishery Science (UiT). She has been working as a social scientist since 2000, with a research focus on issues like rural development, sustainable management of common owned resources, user participation, indigenous rights, adaptive management, interactive governance, legal pluralism, entrepreneurship, and tourism.

Vera Vilhjálmsdóttir is a researcher at the Icelandic Tourism Research Centre. She graduated from Aarhus University in Denmark with a BSc degree in social anthropology and a Master's degree in sustainable heritage management. Vera's main research topics have been cultural tourism, relating to both pop culture and heritage, and travel behaviours of tourists.

Preface

Being Korean-Australian and nearly Norwegian, I believe I have been searching for the meanings of mobilities in myself, a member of us all—humanity. Growing up in South Korea, an interest in different people, their ways of living, and languages has been a major part of my life since my childhood. Later, having studied culture, society, languages, and people's meanings of the world and their connections to outside worlds and how these are expressed have fascinated me. As a sociologist with a literature-studies background and having immigrated twice in my adult life, I became so interested in learning more about Thorstein Veblen especially when I learnt his disciplinary background was both as an economist and sociologist as well as being a Norwegian immigrant family member in the industrialising American society of 18th-19th century.

My interest in learning more about Veblen and his concept of conspicuous consumption further intensified when I moved to Norway from Australia in 2014. Conspicuous consumption, when I first came across the term as a student, sounded like a word from a glossy magazine; it was sensational and eye-catching, which popular magazine journalists are well trained to conjure up. At the same time, I found it a bit odd that it was a scholar from a century ago who came up with this term.

When I moved to Norway in 2014, I had, with little doubt, an etic (Korean-Australian) view of the country. The etic view I held somehow told me that Norwegian culture does not give much attention to something conspicuous, let alone conspicuous consumption. I think this background position of mine may provide the reason why I found it odd that the term was coined by a Norwegian immigrant heritage scholar in America when I encountered the term in the first instance. As the term increasingly became more noticeable in marketing and management literatures, I began to focus my interest on Thorstein Veblen and his conspicuous consumption. While more references to the term were encountered, my curiosity was piqued because of the seemingly mismatching link between a not-very-conspicuous minded Norwegian heritage academic and the coining of the term a century ago, which has become relished in today's academic work in the late industrial world.

As I researched and learnt more about mobilities in the context of a changing Arctic, my impetus to link the subject of Arctic tourism and Asia (i.e. its people, culture, and the ways Asian people connect with the outside world) grew accordingly. Viewing the life of contemporary industrialised societies as an act of consumption is not an alien concept, and mobilities consumption at a global level is accordingly similarly regarded. This is when I saw the essential need for me to understand Veblen's conspicuous consumption idea in his own words rather than being satisfied with accepting and using the notion as presented in published literature.

Having attempted to understand Veblen's conspicuous consumption in his own words, I see that the use of the term in contemporary marketing and management has over-simplified the idea. I give my analysis of the concept itself and the use of the concept in subsequent literatures in Chapter 2 of this book.

When our species of humanity travelled out of Africa some 2 million years ago, populating the entire globe, the mobilities embraced by our ancestors would have been arduous. There was no pathway open for them, no places to rest or sleep, and uncertainty about what to eat along the way. Nonetheless, the mobilities were executed, and our planet began to see and feel the impacts of the expansion of our species. In a way, I see a parallel between the original mobilities some 2 million years ago and the mobilities in today's changing Arctic. What has altered is the developed infrastructures and superstructures for exchanging cultural and environmental experiences, business transactions, distributing costs and benefits of such exchanges, and so forth that enable mobilities to the Arctic. Concurrently, mobilities to the Arctic is occurring at a time of economic and political re-ordering of the world where an Asian century poses a serious challenge to the established thoughts and systems that have solidified over the past nearly 300 years. Mobilities by sapience marked the beginning of human civilisation and prosperity on this planet. The mobilities to the changing Arctic, at least partially, comes from a curiosity to witness a vanishing part of the planet. It depends on us whether this vanishing part of the planet will include our own existence as part of the planet. Being able to comprehend the ways to consume the Arctic in a sustainable way will contribute to continuing our existence here on this planet.

As an early-stage enquiry into Asian mobilities consumption in a fast-changing Arctic, this book takes an Arctic viewpoint. Researchers in Arctic destinations or who are associated with the Arctic through their research interests and/or lived experiences have made this attempt possible. I would like to express my sincere gratitude to the colleagues who saw the value of such research and joined me in this journey. The chapters' contributors and I appreciate the patience and tremendous insights given by Professor Gayle Jennings during the internal proofreading stage of this book project. We hope that the journey will continue to deepen our knowledge for our continued life and culture on this planet.

Acknowledgements

This book project has been made possible with the generous support of the University of the Arctic (UArctic) and the Arctic Five network at UiT the Arctic University of Norway.

Part 1

Setting the scene

Asian mobilities consumption and the Arctic

1 Asia, mobilities, and consumption power and the Arctic

Young-Sook Lee

The Asian century and the changing Arctic: Arctic tourism as the epitome of Arctic politics

In various sectors, the 21st century is considered an Asian century (Kharas & Kohli, 2011). Global financial and political scenes were at the frontline engaging in the Asian-century discussion. When the Asian Development Bank claimed an "Asian century" by 2050, the argument was founded on measurable economic results, i.e. Gross Domestic Product (GDP). Based on projected production, it was forecasted that by 2050, Asia is expected to account for 52% of the world's GDP. Accordingly, this would place Asia as the economic leader of the world as it had been some 300 years ago, before the industrial revolution (Asian Development Bank, 2011). This forecast seems to have been rather modest. Indeed, in 2019, the World Economic Forum confirmed that in 2020 Asia's GDP would already have overtaken the GDP of the rest of the world combined. It further forecasted that by 2030, the region is expected to contribute to roughly 60% of global growth (Yendamuri & Ingilizian, 2019).

The making of an "Asian century" has been resonating in political spheres, particularly over the governance of the Arctic. In 2013, Asian nations, i.e. China, India, Japan, Singapore, and South Korea, have entered the Arctic Council as non-Arctic state observers (The Arctic Council, 2020). Such entry recognises the institutionalisation of Arctic governance as an entity beyond the geographically confined Arctic nations. The global issue of climate change and the interconnected nature of the global economy in the 21st century may have facilitated and strengthened the link between the Arctic and Asia. In an article published in the same year when the Asian countries were accepted as non-Arctic state observers to the Arctic Council, Stokke (2013) confirmed the positive aspects of cooperation between the Arctic and Asia. Having reviewed concerns over Asia's closer involvement in Arctic matters, the paper highlighted the fact that many activities and regulatory bodies, e.g. the United Nations, UN-based climate regime, and the International Maritime Organization (IMO), are indeed of international scale. Such scale is expected to bring out global-level consequences for actions and monitoring systems. With this confirmation of the nature of Arctic

DOI: 10.4324/9781003039518-1

matters being global in their consequences, Asian cooperation has been assessed as a promising one.

Scepticism of Asia's increasingly notable presence in the Arctic is not hard to witness. Following the acceptance of Asian countries into the Arctic Council, concerns over, in particular, growing Chinese influence in Arctic matters augmented among Arctic nations. Taking an Icelandic perspective, Hastings (2014) illustrated that increasing Chinese investments and China's overall overture to the Arctic has been received with caution. In political and scientific realms, Iceland seems to be positively open to cooperation with Asia, but its attitude towards economic collaboration was rather ambivalent. This nuanced stance appears to be due to the imbalanced magnitude in power and influences. As clear evidence of an ambivalent perception to economic cooperation, in 2011, a Chinese tourism investment proposal to construct an ecotourism venue in a peripheral region of Iceland might illustrate a clearer picture. A Chinese investment group, Beijing Zhongkun, and its president Huang Nubo applied for an exemption from laws banning foreign investment in real estate from parties outside the European Economic Area (EEA). The Ministry of Interior in Iceland declined the request (Huijbens & Alessio, 2015). This decision has received much attention across media and political debates along with tourism practitioners and researchers. The attention brought out various issues of Arctic governance, international cooperation, and economic sovereignty that often are expressed through investments such as real estate or tourism engagements. The scepticism seems to continue. In 2019, Babin and Lasserre's (2019) study investigated Western states' perceptions regarding growing Asian involvement in the Arctic. The study reviewed and compared academic publications between Asia and the West, which were combined with reviews of media coverage and interviews. The study argued that the actual power that can be imposed by Asian nations at the policy-making level is considerably little, if any. Thus, Asian involvement should not be viewed with too much caution. It is evident that after some six years of Asian countries' acceptance as part of the Arctic Council, scepticism over Arctic governance and indigenous people's rights appear to remain vivid despite the logical assessment of the studied materials and results.

As the Icelandic experience with Chinese investment attempts to construct an ecotourism venue has shown, tourism in the Arctic has a strong connection to politics. As Asian influences across the globe continue to expand, the tourism sector is not immune to inter-connected global structures, systems, and politics. Rather than being immune to it, the tourism sector might be aptly called an arena where international politics meet and exchange various perspectives and actions. In this changing Arctic, coupled with the re-ordering of global politics, Arctic tourism indeed could well be understood as an epitome of international Arctic politics.

Politics in tourism and mobilities: heightened by horizontal dimensions in a changing Arctic

The expanding Asian influence in political and economic spheres on the global stage is notably represented in Arctic tourism with its unprecedented growth in visitor numbers in Arctic destinations (Runge, Daigle, & Hausner, 2020; Rantala et al., 2019; Lee et al., 2017). Asia's growing power, led by China, coincides with the fast melting of Arctic ice due to climate change. The melting icecaps in the Arctic subsequently bring the world's attention to this fast-changing corner of the world as well as Arctic relevant issues in political spheres. In 1994, when the book dedicated to tourism and politics came out, China was described as a country that risked much of its potential tourism success due to its political instability, i.e. Tiananmen Square (Hall, 1994). Over some 26 years, a change occurred in global tourism that might be described as nothing short of fascinating. At the time of writing, we witness China as a country that employs outbound tourism as its political instrument, such as soft power (Johnson, Xu, & Arlt, 2020) or economic statecraft (Lim, Ferguson, & Bishop, 2020). This change in the tourism politics arena highlights a dimension to which we do not seem to give enough emphasis when evaluating tourism.

Since the focus of attention on tourism and politics began, the multi-scaled nature of tourism governance (Hall, 2007) has been used to effectively describe the layered states of tourism governance. Indeed, the multi-scaled nature of governance in Arctic tourism appears to hold true. What we see in addition to the multi-scaled governance in the context of Arctic tourism may be similarly described as a heightened form to which "multi-sectoral", "multi-disciplinarian", and multi-national" governance were added. In other words, Arctic mobilities are enacted on the accepted tourism governance notion, i.e. multi-layered levels of local, regional, national, and international. While Arctic mobilities are operated and performed according to the global stage, horizontal dimensions that interact with Arctic mobilities may be said to be more powerful in mobilities operation in the Arctic than those in other parts of the world. For example, the transportation sector, a fundamental infrastructure for mobilities, has a profound re-configuration state in the Arctic owing to climate change. The Northern Sea Route (NSR) as a future link between Arctic Europe and Asia is becoming more real than in any other preceding times. This situation, in turn, endows the sector with more political, economic, and environmental powers when it comes to enabling or disabling Arctic mobilities. In this sense, one may contend that a vertical nature in multi-scaled tourism governance in the global scene is met by a horizontally dimensioned nature in the changing Arctic. Indeed, this heightened version of layers and sectors may provide a ground for more nuanced trajectories of Arctic tourism.

Economic and political power is shifting towards Asia (Chhabra et al., 2020), where production, transportation, and consumption occur in a densely populated and strategically aligned fashion. Subsequently, social scientists refreshed their assessment of tourism as a broader and more encompassing notion of mobilities, which further provides Asia with even more assets in their contributions to Arctic tourism. Researchers of the early 21st century began to see that enquiries into tourism studies needed to expand to be more inclusive. This more inclusive version of tourism should consider related transportation systems, political settings that enable or disable the flow of the people and goods and services, and objects of transportation and their cultural domains, among others. Such a view gave rise to a new approach to tourism studies, which was re-configured as mobilities (Sheller & Urry, 2004). Some went as far as calling this new configuration a new paradigm, i.e. the mobilities paradigm (Brembeck, Cochoy, & Moisander, 2015; Cohen & Cohen, 2015). There was also a different view, which asserted that the new approach may not quite qualify as a paradigm as such but still was an important turn in enquiries into travel and tourism (Harrison, 2017).

Debate regarding the new "paradigm" highlights that tourism studies have privileged a Eurocentric, if not primarily a British view during the 20th century (Ryan, 2003). Relatedly, the paradigm/turn needs to incorporate broader environmental, political, cultural, social, and economic contexts in order to gain a better understanding of the phenomenon and its practical operations. A more comprehensive approach to studying tourism has never been more relevant than today in a fast-changing Arctic context. The current volume is an attempt to gain knowledge of consumption by Asian visitors, led by Chinese, of the Arctic where an unprecedented change is occurring. In this book, mobilities, thus, focuses on tourist-related mobilities, while bringing other aspects of mobilities, e.g. environmental, political, and social changes to infrastructures and services that facilitate people and goods to move into broader consideration.

Consumption by Asia and its mobilities towards the Arctic

Concurrent with the debate regarding mobilities, i.e. whether it is a new paradigm or a broader approach to study tourism, the global power of consumption appears to have moved to Asia (Kharas, 2017). With the re-ordering of power and an economic centre gravitating towards Asia, researchers investigate various links that Asian consumption patterns have in broader sectors. Confirming a shift of the consumption centre to Asia, Kim (2017) found that consumption-driven economic growth is a distinct Asian pattern of economic growth. The study further pointed out that while consumption-driven economic growth is evident in Asian countries, investment, export, and savings-driven economic models, conventional factors for economic growth, still have significantly positive effects in Asian societies. With the expanding consumption power of Asia, Asian tourists with their ways of consuming nature, culture, and materials have already arrived at Arctic destinations with a fast growth rate in recent years. The arrival has been facilitated by climate change and a re-ordering of international political powers in the Arctic.

Asian mobilities towards the Arctic have for a long time been in the process of making when we consider the Northern Sea Route (NSR). The concept of the NSR or Northeast Passage as called by Western Europeans arose in the 16th century, according to Armstrong's (1992) work on the history of the NSR. Throughout the ensuing centuries, there have been attempts to find a route in the Northern Sea by traders and explorers. The final and actual discovery of the route is attributed to a land-based expedition led by Baron F.P. Wrangel in 1821–1824. As a contested space, the NSR attracted attention from marine policy and logistics/shipping sectors regarding its viability as a sustainable trade route between Europe and Asia. Until recent years, studies have been not without reservations or only with certain conditions to argue the NSR's economic viability for the shipping industry (Liu & Kronbak, 2010; Zhang, Meng, & Zhang, 2016). While the economic viability of the NSR has been considered, there has been a distinct view that the shifting of economic geography or geopolitical changes between Europe and Asia might play important roles in the development of the NSR (Blunden, 2012). Accordingly, broader environmental issues and the post-2020 sulphur regulations by International Maritime Organization [*sic*] (IMO) are considered in more recent investigations into the profitability of the NSR (Wang, Ding, Gong, Wang, Wang, & Huang, 2020; Keltto & Woo, 2020). The post-2020 sulphur regulations refer to the IMO's adoption from January 2020 to reduce the sulphur content to 0.5% from the previous 3.5% for shipping (IMO, 2020). It is apparent that the longstanding making of Asian mobilities towards the Arctic is placed within the broader economic and environmental spheres of the Arctic.

Asian mobilities and their consumption in a changing Arctic

Research interests into Asian mobilities have been expressed previously in tourism studies. For example, a set of four papers presented Asian mobilities accompanied by two analytical editorial notes. Identifying the Asian mobilities characteristics, i.e. relationalities, re-inventions, and resistance, Lin and Yeoh (2016) framed the enquires into Asian mobilities, which are enacted from its own Asian epistemological and ontological standpoints rather than a mere reproduction of Western-centric phenomena. The collection of papers presents accounts of a demand for credibility in Singapore's newly opened Casino, a demonstration of a new mobile identity (Zhang & Yeoh, 2016), the making of airspace founded on the Singaporean experience (Lin, 2016), Mumbai's taxi-trade being modernised with a Singapore-style mobility (Bedi, 2016), and geographic margins not being considered as part of the state based on a case from Karachi, Pakistan (Anwar, 2016). Analytically diagnosing the current state of mobilities study as being West/Anglo/Eurocentric, Cresswell (2016) highlighted the importance of researching mobilities that are being enacted in Asia. Another piece of research on Asian mobilities illustrated how Sri Lanka has experienced changes in cultural values through mobilities. In research on mobilities in Sri Lanka, Ranasinghe and Li (2017) conducted 27 un-structured in-depth interviews and recounted that dimensions of mobilities in Sri Lankan society were attributed to the emerging new cultural hegemony of China, a new turn from its British colonial history.

The earlier work on Asian mobilities has focused on Asian epistemologies enacted in Asian locations. Comparatively, the current book focuses on Asian mobilities with its epistemological grounding being performed in Arctic destinations. Accordingly, the central point of the stories in this book is expected to privilege Asian worldviews as they interact and are performed in Arctic destinations.

Our current understanding of Asian consumption appears to be closely linked to and understood as conspicuous consumption, as conceptualised and introduced by Thorstein Veblen in 19th-century America. Indeed, researchers identify Asian forms of consumption as good examples of Veblen's 19th-century notion. Some have tried to make a link to Asian cultural philosophies in order to explain why the 19th-century American defined concept aptly explains 21st-century Asian forms of consumption. For example, Podoshen, Lu, and Zhang (2011) suggest that the Chinese concept of face (*mien-tsu*) facilitates overtly conspicuous consumption. Similarly, other studies identify the seemingly delayed yet nearly identical patterns of conspicuous consumption by Asians and, in some cases, the level of conspicuousness displayed by Asians exceeding those of Westerners (Wong & Ahuvia, 1998; Park, Rabolt, & Jeon, 2008). Still, a cautious view warns that defining the Asian way of consumption as a static consumption risks limiting our understanding of Asian historic and cultural bodies of consumption (Francks, 2009).

Concurrent with the attempts to understand Asian consumption related to Veblen's conspicuous consumption, a specific mode of consumption, i.e. luxury consumption by Asians has been closely investigated. Defining luxury consumption as being the opposite of necessity (Mo, 2020) or attributing its impetus to consumers' specific motivations (Zhan & He, 2012), research into luxury consumption invariably links its theoretical basis to Veblen's work in 1899, *The theory of leisure class*, where the notion of conspicuous consumption was conceptualised and introduced for the first time. In tandem with the research focus given to Asia's luxury consumption, luxury management education programmes at higher education institutions have been on offer across Europe and Asia and how to manage luxury (Som & Blanckaert, 2015) has become an important subject to cover.

As global mobilities and consumption are increasingly connected with broader settings of society, one is left to wonder about the adequacy of uncritically adopting the 19th-century Norwegian American theory to explain 21st-century Asian mobilities towards the Arctic. This enquiry seems particularly pertinent at a time when the theoretical divide, even within Western-centric enquiries, is apparent (Sharpley, 2020). One sees an essential need to stop and reflect on the ways in which we observe, theorise, and accept. This volume exerts our research attentions to better understand 21st-century Asian mobilities consumption in Arctic destinations, evaluating their adequacy and how we might more properly use Veblen's conspicuous consumption.

Veblen's conspicuous consumption for Asian mobilities consumption of the Arctic?

When we consider the adequacy of Veblen's conspicuous consumption notion to explain Asian mobilities consumption in the 21st-century Arctic, it makes sense to try to understand how Veblen and his consumption idea have been understood in other fields of study. Economists defined Veblen effects as "a willingness to pay a higher price for a functionally equivalent good…" (Bagwell et al., 1996, p. 350). This definition led to further understanding consumption by individuals. That is, if the Veblen effect is in operation, variations in consumption levels will be related not to individuals' own incomes but to those of better off groups (Bowles & Park, 2005). The apparent existence of the Veblen effect has been traced in leisure studies. In his study on income inequality and leisure, Veal (2016, p. 237) contended that "the proposition that people's consumption patterns partly reflect not just their own income but the consumption patterns of higher income groups".

Both consumptions patterns in mainstream economics and leisure contexts might be attributed to Veblen's pecuniary emulation, which betrays people's desire to be perceived as more economically prosperous than they are by other members in a given society. This trait of pecuniary emulation then leads to conspicuous consumption or in some recent cases, luxury consumption. At the time of growing Asian mobilities towards Arctic destinations and their form of consumption that is expected to increase accordingly, it is high time to make a dedicated enquiry into conspicuous consumption related to Asian mobilities in the Arctic. The following chapter presents a theoretical evaluation of Veblen's concept of conspicuous consumption related to his leisure class.

Organisation of the book

At a global level, Arctic tourism is arguably an epitome of Arctic politics. Globe-gripping climate change heightens the complexity of Arctic mobilities. Arctic mobilities encompass the interactions between vertical scales and horizontal dimensions: local, regional, national, and/or international level (vertical) scales interacting with multi-sectoral, multi-national, multi-cultural, and multi-disciplinary (horizontal) dimensions. This interaction creates what may be termed as Arctic mobilities characteristics.

This book is thus organised into five parts where the vertical and horizontal aspects meet, and their interactions are traceable from the empirical drawings in the individual chapters. The first part of the book opens with the setting of a scene for Asian mobilities consumption in the Arctic at the onset of an Asian century as described in this current Chapter 1. The chapter also draws attention to Veblen's conspicuous consumption as used in depicting and seemingly explaining Asian forms of consumption. With the introduction of Asian mobilities in the Arctic and Veblen's theory, Chapter 2 critically evaluates current understandings and use of Veblen's conspicuous consumption. The theoretical evaluation of the concept brings out two forms of conspicuous consumption: a) conspicuous consumption as per Veblen's time and space; and b) conspicuous consumption forecasted by Veblen, specifically, beyond Veblen's time and space. The chapter continues to investigate possible interaction trajectories between the second form of conspicuous consumption and the enduring East Asian cultural philosophy, i.e. the five Confucian constants.

The second part of the book presents three chapters that illustrate Asian mobilities consumption as an extension of Arctic international politics. Björner, Dybtsyna, and Lundberg, in Chapter 3, recount the ways in which tourism destination and branding is portrayed in Swedish, Norwegian, and Finnish Arctic strategies and how those portrayals might correspond to the Chinese Arctic strategy.

In Chapter 4, Kobzeva presents Chinse tourism in the Russian Arctic and the Far East. In so doing, the chapter considers the role of Chinese tourism in those regions, types of consumption undertaken by visiting Chinese tourists, and the key challenges that Chinese tourism faces. The next chapter by Reigner, Hovgaard, and Gaini explains the current states of tourism in Greenland and Faroe Islands, one of the fastest growing destinations in the Arctic region. These destinations are sought after by newer and more off-the-beaten path finding adventurers. The authors give a convincing account of the reason why and how the development of tourism in Greenland and the Faroe Islands needs to provide a shared benefit for visitors and the local populations, particularly in a post-Covid era.

The third part of the volume is themed around the transportation sector. Two chapters demonstrate the significance of the cruise industry embedded in the changing social and political settings of the Arctic. In Chapter 6, Eikeland and Chiesa show the way in which renewable energy-based cruise tourism can be achieved by focusing on the city of Tromsø, Northern Norway. Approached from a sustainable engineering perspective, the chapter presents the current state of cruise tourism, its energy consumption as well as the technical feasibility of transitioning to a renewable energy-based cruise tourism industry. Chapter 7 by Ojala, Räikkönen, and Kiiski evaluates the feasibility of the Northeast passage as a tourism transport route between China and the Arctic. Based on traffic statistics regarding the passage and information from tour operators, the chapter considers a way to yield sustainable tourism traffic in a post-Covid era.

The following part presents six chapters dealing with ethnicity, culture, and history, and Asian tourists' Arctic experiences in a changing Arctic. This fourth part starts with Kotašková and Jelínek's critical account of the notion of Asian ethnicity in an attempt to understand Asian mobilities consumption in the Arctic. Exploring the notion of Asian ethnicity through an ethnography of guided tours in Svalbard, the chapter challenges rigid categorisations of ethnicity and highlights the significance of individual performances in the Arctic tour setting. Chapter 9 by Mathisen and Søreng introduces Thai mobilities and consumption of North Cape in Norway. Tracing King Chulalongkorn of Siam's visit to North Cape in 1907, the chapter focuses on the meaning of The Stone, where the King left his signature, for today's Thai mobilities and consumption of that place. Next, in the following chapter, Cheung recounts his lived experiences of being an Arctic adventure guide for Chinese tourists. Employing an analytic autoethnographic approach, Cheung argues that a more dynamic understanding of Chinese adventure mobilities and consumption is essential. Sokolíčková, in Chapter 11, presents Chinese tourism in Svalbard. Using an anthropological approach to understand the meaning of Chinese tourism and consumption in Svalbard at the time of a changing and disappearing Arctic provides valuable insight into the phenomenon. In recent years, the fast-growing trend of Asian tourist numbers has been strongly felt in Iceland. This trend is investigated and presented by Vilhjálmsdóttir, Gunnarsdóttir, and Black in Chapter 12. Based on interviews with tour guides who have extensive experiences of hosting Chinese tourists in

Iceland, the chapter proposes key elements in Chinese tourism and consumption in Iceland. The last chapter in this part provides an architect's viewpoint of urbanising the Arctic in relation to growing Asian mobilities and space consumption in the Arctic region. Departing from an urbanising Arctic and the design of winter cities, Chapman, Lee, and Larsson argue that Asian mobilities consumption may be seen as the outcome of interactions among the Arctic's built environment, individuals, and the changing climate. Recommendations are given regarding Arctic urban design with an aim of reaching sustainable Asian mobilities consumption of spaces in the urbanising Arctic.

The final part consists of the last chapter of the book and revisits the horizontal and vertical interactions of Arctic mobilities consumption in conjunction with Veblen's conspicuous consumption and the Confucian five constants. Capturing noticeable points in the observed Asian consumption of the mobilities to the Arctic, the chapter proposes future directions for research.

References

Anwar, N. H. (2016). Asian mobilities and state governance at the geographic margins: Geopolitics and oil tales from Karachi to Taftan. *Environment & Planning A*, *48*(6), 1047–1063.

Armstrong, T. (1992). A history of the Northern Sea Route. *International Challenges*, *12*(1), 34–42.

Asian Development Bank. (2011). *Asia 2050: Realizing the Asian century*. Manila: Asian Development Bank.

Babin, J., & Lasserre, F. (2019). Asian states at the Arctic Council: Perceptions in Western States. *Polar Geography*, *42*(3), 145–159. doi:10.1080/1088937X.2019.1578290

Bagwell, L. S., & Bernheim, B. D. (1996). Veblen effects in a theory of conspicuous consumption. *American Economic Review*, *86*(3), 349–373.

Bedi, T. (2016). Mimicry, friction and trans-urban imaginaries: Mumbai taxis/Singapore-style. *Environment and Planning A*, *48*(6), 1012–1029.

Blunden, M. (2012). Geopolitics and the Northern Sea Route. *International Affairs*, *88*(1), 115–129. doi:10.1111/j.1468-2346.2012.01060.x

Bowles, S., & Park, Y. (2005). Emulation, inequality, and work hours: Was Thorsten Veblen right? *The Economic Journal*, *115*, F397–F412.

Brembeck, H., Cochoy, F., & Moisander, J. (2015). Moving consumption. *Consumption Markets & Culture*, *18*(1), 1–9. doi:10.1080/10253866.2014.899218

Chhabra, T., Rush, D., Ryan, H., & Kimball, E. (2020). Global china: Great powers. Brookings report series. *Global China*. Retrieved from https://www.brookings.edu/wp-content/uploads/2020/02/FP_20200826_great_powers_chapeau.pdf

Cohen, E., & Cohen, S. A. (2015). Beyond Eurocentrism in tourism: A paradigm shift to mobilities. *Tourism Recreation Research*, *40*(2), 157–168. doi:10.1080/02508281.2015.1039331

Cresswell, T. (2016). Afterword – Asian mobilities/Asian frictions. *Environment and Planning A*, *48*, 1082–1086.

Francks, P. (2009). Inconspicuous consumption: Sake, beer, and the birth of the consumer in Japan. *The Journal of Asian Studies*, *68*(1), 135–164.

Hall, C. M. (1994). *Tourism and politics: Policy, power and place*. Chichester: Wiley.

Hall, C. M. (2007). Tourism, governance and the (mis-)location of power. In A. Church & T. Coles (Eds.), *Tourism, power and space* (pp. 247–269). Abingdon: Routledge.

Harrison, D. (2017). Tourists, mobilities and paradigms. *Tourism Management*, *63*, 329–337.

Hastings, J. G. (2014). The rise of Asia in a changing Arctic: A view from Iceland. *Polar Geography*, *37*(3), 215–233. doi:10.1080/1088937X.2014.934315

Huijbens, E. H., & Alessio, D. (2015). Arctic 'concessions' and icebreaker diplomacy? Chinese tourism development in Iceland. *Current Issues in Tourism*, *18*(5), 433–449. doi:10.1080/13683500.2013.837867

International Maritime Organization. (2020). Sulphur 2020 – Cutting sulphur oxide emissions. Retrieved from http://www.imo.org/en/MediaCentre/HotTopics/Pages/Sulphur-2020.aspx

Johnson, P. C., Xu, H., & Arlt, W. G. (2020). Outbound Chinese tourism: Looking back and looking forward. *Journal of Policy Research in Tourism, Leisure and Events*, *12*(1), 1–7. doi:10.1080/19407963.2018.1505098

Keltto, T., & Woo, S-H. (2020). Profitability of the Northern Sea Route for liquid bulk shipping under post 2020 sulphur regulations. *The International Journal of Logistics Management*, *31*(2), 313–332. doi: 10.1108/IJLM-12-2018-0314

Kharas, H. (2017). The unprecedented expansion of the global middle class: An update. Global economy and development at Brookings. Retrieved from https://www.brookings.edu/wp-content/uploads/2017/02/global_20170228_global-middle-class.pdf

Kharas, H., & Kohli, H. A. (2011). Asia in the global economy in 2050: The Asian century. In H. Kohli, A. Sharma & A. Sood (Eds.), *Asia 2050: Realizing the Asian century* (pp. 43–48). New Delhi: Sage.

Kim, H. (2017). The effect of consumption on economic growth in Asia. *Journal of Global Economics*, *5*, 259. doi:10.4172/2375-4389.1000259

Lee, Y.-S., Weaver, D., & Prebensen, N. K. (2017). *Arctic tourism experiences: Production, consumption and sustainability*. Wallingford: CAB International.

Lim, D. J., Ferguson, V., & Bishop, R. (2020). Chinese outbound tourism as an instrument of economic statecraft. *Journal of Contemporary China*, 1–18. doi:10.1080/10670564.2020.1744390

Lin, W. (2016). Drawing lines in the sky: The emotional labours of airspace production. *Environment and Planning A*, *48*(6), 1030–1046.

Lin, W., & Yeoh, B. (2016). Moving in relations to Asia: The politics and practices of mobility. *Environment and Planning A*, *48*(6), 1004–1011.

Liu, M., & Kronbak, J. (2010). The potential economic viability of using the Northern Sea Route (NSR) as an alternative route between Asia and Europe. *Journal of Transport Geography*, *18*(3), 434–444. doi:10.1016/j.jtrangeo.2009.08.004

Mo, T. (2020). "Income vs. education" revisited – The roles of "family face" and gender in Chinese consumers' luxury consumption. *Asia Pacific Journal of Marketing and Logistics*, *33*(4), 1052–1070. doi:10.1108/APJML-12-2019-0733

Park, H-J.; Rabolt, N. J., & Jeon, K. S. (2008). Purchasing global luxury brands among young Korean consumers. *Journal of Fashion Marketing and Management*, *12*(2), 244–259.

Podoshen, J. S., Li, L., & Zhang, J. (2011). Materialism and conspicuous consumption in China: A cross-cultural examination. *International Journal of Consumer Studies*, *35*(1), 17–25.

Ranasinghe, R., & Cheng, L. (2017). Dimensions of mobilities, tourism and transition of cultural hegemony: A qualitative inquiry from Sri Lanka. *International Journal of Tourism Anthropology*, *6*(1), 21–40. Retrieved from http://www.inderscience.com/info/ingeneral/forthcoming.php?jcode=IJTA

Rantala, O., de la Barre, S., Granås, B., Jóhannesson, G. T., Müller, D. K., Saarinen, J., … Niskala., M. (2019). *Arctic tourism in times of change: Seasonality. Tema Nord 2019*, 528. Copenhagen: Nordic Council of Ministers.

Runge, C. A., Daigle, R. M., & Hausner, V. H. (2020). Quantifying tourism booms and the increasing footprint in the Arctic with social media data. *PLoS ONE*, *15*(1): e0227189. doi:10.1371/journal.pone.0227189

Ryan, C. (2003). *Recreational tourism: Demand and impacts*. Clevedon: Channel View Publications.

Sharpley, R. (2020). Tourism, sustainable development and the theoretical divide: 20 years on. *Journal of Sustainable Tourism*, *28*(11), 1932–1946. doi:10.1080/09669582.2020.1779732.

Sheller, M., & Urry, J. (Eds.). (2004). *Tourism mobilities: Places to play, places in play*. London: Routledge.

Som, A., & Blanckaert, C. (2015). *The road to luxury*. Singapore: John Wiley & Sons.

Stokke, O. S. (2013). The promise of involvement: Asia in the Arctic. *Strategic Analysis*, *37*(4), 474–479. doi:10.1080/09700161.2013.802520

The Arctic Council. (2020). Non-Arctic states. Retrieved from https://arctic-council.org/en/about/observers/non-arctic-states/

Veal, A. J. (2016). Leisure, income inequality and the Veblen effect: Cross- national analysis of leisure time and sport and cultural activity. *Leisure Studies*, *35*(2), 215–240. doi:10.1080/02614367.2015.1036104

Wang, D., Ding, R., Gong, Y., Wang, R., Wang, J., & Huang, X. (2020). Feasibility of the Northern Sea Route for oil shipping from the economic and environmental perspective and its influence on China's oil imports. *Marine Policy*, *118*(2020), 104006. doi:10.1016/j.marpol.2020.104006

Wong, N. Y., & Ahuvia, A. C. (1998). Personal taste and family face: Luxury consumption in Confucian and Western societies. *Psychology and Marketing*, *15*(August), 423–441.

Yendamuri, P., & Ingilizian, Z. (2019). In 2020 Asia will have the world's largest GDP. Here's what that means. *World Economic Forum Annual Meeting*. Retrieved from https://www.weforum.org/agenda/2019/12/asia-economic-growth/

Zhan, L., & He, Y. (2012). Understanding luxury consumption in China: Consumer perceptions of best-known brands. *Journal of Business Research*, *65*(10), 1452–1460. doi:10.1016/j.jbusres.2011.10.011

Zhang, J., & Yeoh, B. S. (2016). Harnessing exception: Mobilities, credibility, and the casino. *Environment and Planning A*, *48*, 1064–1081.

Zhang, Y., Meng, Q., & Zhang, L. (2016). Is the Northern Sea Route attractive to shipping companies? Some insights from recent ship traffic data. *Marine Policy*, *73*, 53–60.

2 A critique on the non-critical application of Veblen's conspicuous consumption

A perspective on Asian mobilities in the Arctic

Young-Sook Lee

Introduction

Where Asian consumption is concerned, academic literature in marketing and tourism studies convinces us that Thorstein Veblen's conspicuous consumption notion explains Asian consumption quite satisfactorily (Li, Guo, Zhang, & Sun, 2019; Phau & Teah, 2009; Zakaria, Wan-Ismail, & Abdul-Talib, 2020; Park & Reisinger, 2009). Superficially, the concept seems to aptly describe and explain the ways in which Asians consume. On a closer inspection, however, one is left to wonder if the application of the 19th-century Norwegian American economist and sociologist's idea regarding 21st-century Asian consumption is indeed adequate.

With a focus on Asian mobilities consumption in a changing Arctic in the 21st century, this chapter presents a theoretical framework based on Veblen's concept of conspicuous consumption related to Asian mobilities consumption in the Arctic. The chapter, accordingly, serves as the theoretical grounding upon which a number of relevant points are revisited in several other chapters in this book.

This chapter has three parts. The first section provides insights into Veblen as a person, a scholar, and an agent of his era of human history. The second section reviews his theory of leisure, the leisure class, and conspicuous consumption using Veblen's own terms of his era. The review on leisure and its class in Veblen's epoch provides background to the notion of conspicuous consumption because the leisure concept gives rise to the very idea of conspicuous consumption. The final section of the chapter presents a framework that incorporates Veblen's conspicuous consumption and fundamental elements in Asian mobilities consumption informed by the five Confucian constants, a fundamental base in East Asian worldviews, which extend into the current 21st century.

Veblen: the person, scholar, and agent of his era

Veblen reflects certain traits of a second-generation Norwegian immigrant to America as a person, and, being both economist and sociologist, as a scholar

DOI: 10.4324/9781003039518-2

in late 19th and early 20th-century America. Born to an immigrant family from Norway in Manitowic County of Wisconsin, Thorstein Veblen's (1857–1929) view on human history is not surprisingly grounded in a Euro/Western epistemology. In using the term, Euro/Western, the current author's intention is not to homogenise European and Western thoughts on human understandings of knowledge. Rather, the intention is to highlight that his worldview appears to have been drawn from the dominant paradigmatic strains in the modern world of sciences. Although his upbringing in American society at the time may be said to not be mainstream (McCormick, 2011) *per se*, however, as an immigrant family from a Nordic country, his values and religious background might be said to be less alienated from 19th-century American society compared to those from different parts of the globe at the time, e.g. Asia or Africa.

Accordingly, his worldview places the progress of history into a linear model, particularly his assessment of individual ownership, which gives rise to the leisure class (see Introduction chapter in Veblen, 1899/2009). That is, as time progresses, human society and economic and social phases become better or progressive towards a more correct way. It is in this light that he was one of the first to apply the biological theory of Darwinism to social institution and structure, namely: Social Darwinism (Hodgson, 2008). Adopting the view of the survival and development of humans in social settings from biology, Social Darwinism subscribes to the natural law of "survival of fittest". Applying this physical natural law into social domains has been deemed controversial and the idea has been disputed among researchers (Rogers, 1972). Veblen's exceptional capacity as a thinker, who saw the link between the natural law and human society, may also be linked to his work: *The Instinct of Workmanship* (1914). According to Plotkin (2017), the work has been evaluated as the most valuable study by Veblen. It should be noted that the author focuses on Veblen's exceptional capacity as a thinker in this chapter, while aware but not delving into further on some of his other controversial traits, e.g. his sorry plight of women.

Veblen ascribed the origins of institutions to learned habits and ultimately to innate instincts, which provide a set of basic drives of human action, in the context of particular material conditions (Edgell, 1975 cited in Corde, 2005, p. 1). This summation of Veblen's conception of instinct of workmanship clearly highlights the coordinating state between biological (to innate instincts) and social (institutions, learned habits) elements in humanity. Indeed, this state of harmonisation between biological and psychological dimensions of being human is clearly presented in Veblen's own assertion on the concept of "instinct" as highlighted in Waller's (2017) review. Delving into the very core of Veblen's concept of instinct, Waller presents an extract, containing the following text:

> In these biological sciences interest is centering [*sic*] not on the question of what activities may be set down in innate propensity or predisposition at large, but rather on the determination of the irreducible psychological—and indeed, physiological—elements that go to make up animal behavior [*sic*].

> For this purpose "instinct" is a concept of too lax and shifty a definition to meet the demands of exact biological science.
>
> (Veblen, 1890/ 1990 cited in Waller, 2017, p. 41)

Veblen's discomfort with the biologists' approach to understanding human existence is clearly expressed in the above extract. He fiercely criticises the natural scientists' interests being focussed on minimising the psychological and physiological coordination in making up what being human is. In this sense, Waller's review of the essence of Veblen's instinct, which might be summarised at least partially as a non-static, ever-changing, and adapting state that provides foundations for individual proclivity to human actions within the context of material conditions, is indeed comprehensive and profound.

Veblen's struggle to show his peers the inherent link between biological and psychological dimensions in human existence and behaviour is arguably a sign of him being a social agent of the era. One can hardly deny that applying biological theory to social sciences reflects the awakening of the modern scientific era with the realisation that a human *is* both a biological and cognitive being. In the late 19th-century epoch, when Veblen's conspicuous consumption was conceived, inquiries into what constituted being a human were vigorous. This was a time when the fundamental base for modern psychology was being established by Freud (1856–1939) and Jung (1875–1961). Both identified the inherent links between physiological and psychological dimensions of being human. The realisation of biological and psychological connectedness extended beyond scientific realms. Indeed, the general public seemed to have had just as much fascination as the scientists into inquiries into the nature of being human and human behaviour. It may not have been entirely coincidental that *The Strange Case of Dr Jekyll and Mr Hyde* by Robert Louis Stevenson in 1886 received great interest from the general public. The novel tells a story of an eminent scientist, who temporarily succeeded in separating evil and good in himself thus producing two different persons from him: one pure evil and one pure good. The story goes on to tell the eventual failure of such an experiment and the subsequent demise of the scientist. Given the temporal context of Veblen's theorisation of consumption, he could aptly be described as an agent of his era in human history.

Leisure and leisure class by Veblen

The concept of leisure and leisure class in human society is the fundamental backdrop in Veblen's conspicuous consumption. According to Veblen, what led to human society's leisure and its class was individual ownership. From a primitive accumulation of individual ownership, human society moves to efficient industry-based accumulation. With the industry-based accumulation, "excess" for personal individual consumption of goods inevitably appears, which is deemed good and thus sought by esteemed parts of society. Over time, in Veblen's worldview, human society progresses or becomes better. Further, "peaceable states of human history", as periodised by Veblen and in which primitive cultures are found, were not deemed worthy of consideration regarding individual ownership or accumulation, as there was no fighting among individuals to own more than other members of a society. When individual ownership existed, its existence and effect were so minimal it did not provide much difference in the overall context. This peaceable stage, however, changes over time where Social Darwinism, or the "survival of fittest" becomes the habitus in human society. The "survival of fittest" paradigm in human society brings on the advent of the predatory stage in history where power and class division appear apparent. At this stage of human history, predatory and exploitive employment (i.e. work that is not essential for human existence such as administrative work for a society, artistic work, or religious roles) is regarded worthy and prestigious, while industrial employment (i.e. work required for basic human existence such as hunting, cooking, cleaning) is shameful and drudgery (see Chapter III in Veblen, 1899 for further descriptions). Eventually, the goal of accumulating excess is to obtain a higher ranking in society.

Veblen's concept of leisure and its class appears from the predatory or exploitive stage of human history. What needs to be noted is that his notion of leisure is very different from the way leisure is viewed in the 20th and 21st centuries. Indeed, in Veblen's words, leisure means an absence of productive work (industrial employment) that is essential for human existence. His argument regarding the meaning of leisure is as follows:

> the term "leisure" as here used, does not connote indolence or quiescence. What it connotes is non-productive consumption of time. Time is consumed non-productively (1) from a sense of the unworthiness of productive work, and (2) as an evidence of pecuniary ability to afford a life of idleness.
>
> (Veblen, 1899/1990, p. 33)

To Veblen, consuming time non-productively expresses a person's disdain of productive work. Such disdain, subsequently, displays one's pecuniary ability. Wasting time on non-productive work is the essence of Veblen's leisure.

Interestingly, Veblen refines what he means by waste and further explains that the waste to which he refers is macroeconomically viewed. Describing the term "waste" in the process of enacting conspicuous leisure and consumption of time and material goods, Veblen says:

> In the view of economic theory the expenditure in question is no more and no less legitimate than any other expenditure. It is here called "waste" because this expenditure does not serve human life or human well-being **on the whole** (emphasis added), not because it is waste or misdirection of effort or expenditure as viewed from the standpoint of the individual consumer who chooses it.
>
> (Veblen, 1890/ 1990, p. 67)

Veblen makes it very clear here that what is regarded as waste on the whole may not be regarded similarly by the individual who performs the act of waste. This is the pivotal point where we can appreciate Veblen as an economist and sociologist at the same time. While the economic assessment of the use of time and material goods evaluate the act as wasteful, sociological evaluation may be bringing out the very opposite outcome.

With the conceptualisation of leisure and its class established, the activities engaged by the leisure class demonstrate the concept's characteristics, i.e. non-productive and wasteful. Manners and decorum, fluency in certain obscure languages, for example, become the sign of the person's ability to waste time and material goods. In some cases, vicarious leisure may bring even heightened glory of the "master's" pecuniary ability. That is, if a servant or more than one engages in certain non-productive and wasteful activities in order to serve the master appropriately, this vicarious leisure by servants displays yet a higher level of pecuniarily excessive state of the master. Vicarious leisure consumption is described in more detail in the following part as an act of conspicuous consumption.

Conspicuous consumption by Veblen

Conspicuous consumption as defined by Veblen is an integral part of the leisure domain in his era of the late 19th and early 20th centuries. That is, leisure was defined as time wasting and material goods accumulation in order to demonstrate that one does not need to engage in life-sustaining essential work for her/his own livelihood. Such notions encompass the entire parameters of conspicuous consumption. The fundamental rationale for conspicuous consumption is premised on an excessive surplus being used non-productively and wastefully. These are key parameters of leisure in Veblen's epoch. Accordingly, while enacting Veblen-defined leisure, certain wastes were deemed more wasteful than others, subsequently endowing the performer with more prestige and honour, as expressed clearly here:

> Throughout the entire evolution of conspicuous expenditure, whether of goods or of services or human life, runs the obvious implication that in order to effectually mend the consumer's good fame it must be an expenditure of superfluities. In order to be reputable it must be wasteful.
>
> (Veblen, 1890/ 1990, p. 67)

In an effort to display one's ability to waste at a higher level than others, different layers of conspicuous consumption are employed, i.e. vicarious consumption performed by friends, competitors, wives, and servants:

> Conspicuous consumption of valuable goods is a means of reputability to the gentleman [*sic*] of leisure. As wealth accumulates on his hands, his own unaided effort will not avail to sufficiently put his opulence in evidence by this method. The aid of friend and competitors is therefore brought in by resorting to the giving of valuable presents and expensive feasts and entertainments. ... he consumes vicariously ...

> As wealth accumulates, the leisure class develops further in function and structure, and there arises a differentiation within the class. There is a more or less elaborate system of rank and grades.
>
> Throughout this graduated scheme of vicarious leisure and vicarious consumption the rule holds that these offices must be performed in some such manner, or under some such circumstance or insignia, as shall point plainly to the master to whom this leisure or consumption repute of right inures.
>
> (Veblen, 1890/ 1990, p. 53–55)

Such unproductive consumption of materials goods, time, and money is, indeed, honourable in Veblen's conceptualisation of conspicuous consumption. As hinted in the above extract, the ranked and graduated scheme of vicarious consumption can achieve the best result when performed in certain correct manners. For a sophisticated conspicuous consumption, one should be able to display a refined way of waste. Refined ways of wasting pecuniary resources are determined by habitus, which establishes itself over a long period of time in a society. Veblen differentiates honourable waste from some stultification this way:

> In order to avoid stultification he must also cultivate his tastes, for it now becomes incumbent on him to discriminate with some nicety between the noble and the ignoble in consumable goods. He becomes a connoisseur in creditable viands of various degrees of merit, in many beverages and trinkets, in seemly apparel and architecture, in weapons, games, dancers, and the narcotics. This cultivation of æsthetic faculty requires time and application, and the demands made upon the gentleman in this direction therefore tend to change his life of leisure into a more or less arduous application to the business of *learning how to live a life of ostensible leisure in a becoming way* (emphasis added).
>
> (Veblen, 1890/ 1990, p. 53)

Veblen's conspicuous consumption notion is located as a part in a bigger arena of leisure in his era. In his conceptualisation, accordingly, leisure and consumption reside separately. The separation as a concept between leisure and consumption was a distinctive feature to Veblen. So clear was the conceptual separation between the two that Veblen predicted that the domain of leisure would lose its ground to that of consumption in the future in this way:

> Leisure might then be expected gradually to yield ground and tend to obsolesce as the economic development goes forward, and the community increases in size; while the conspicuous consumption of goods should gradually gain in importance, both absolutely and relatively, until it had absorbed all the available product, leaving nothing over beyond a bare livelihood.
>
> (Veblen, 1890/ 1990, p. 63–64)

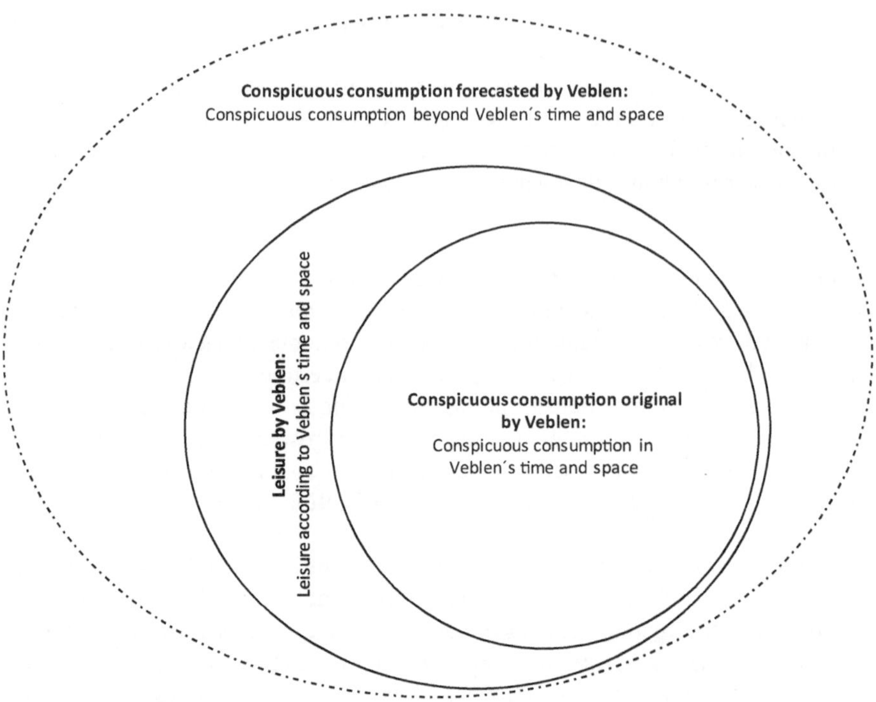

Figure 2.1 Veblen's conspicuous consumption in two forms: in and beyond his time and space

Appreciating the 21st-century consumption-driven global society, one finds little ground to dispute what Veblen forecasted in 1899. His reasoning for consumption overtaking leisure, and further extending far beyond to the leisure domain, rests on the simple social fact that wasting is more efficient to demonstrate one's excessive wealth than leisure, albeit originally it did give rise to the concept of conspicuous consumption. It is obvious from a critical reading of Veblen's conceptualisation of leisure, its class, and conspicuous consumption, that there are two forms of conspicuous consumption in his thesis, as depicted in Figure 2.1. Veblen's conspicuous consumption in two forms: in and beyond his time and space.

Veblen correctly predicted that consumption would take over leisure in future because conspicuous consumption was far more effective than conspicuous leisure in displaying the person's level of excessive wealth. Indeed, we have seen a rise in consumption as the identity-defining agent in human societies throughout the 20th century (see Belk, 1988; Taylor & Strutton, 2016 for more discussions). Veblen's original conspicuous consumption in his adopted homeland of late 19th-century America has now expanded to the global stage. In the 21st century, we need to be clear that we are dealing with Veblen's futuristic conspicuous consumption rather than its original form. However, the adoptions of the notion in literature to date do not discern the two forms of conspicuous consumption. Failing to see the two forms of conspicuous consumption, application of his work remains uncritical and limited to the original form of conspicuous consumption associated with his time and space.

Having delineated the two forms of conspicuous consumption in Veblen's thesis, the future form of conspicuous consumption is the focus of the current chapter's critique. The reason for highlighting the future form of conspicuous consumption is because 21st-century conspicuous consumption is performed at the international level, with the biggest consumers being beyond Veblen's leisure-notion boundaries of 19th-century America. In the early 21st century, the biggest consumers in international tourism are the Chinese (UNWTO, 2018). Beyond the tourism and mobilities sector, one quarter of global consumers are of Confucian heritage (Park & Cho, 1995), with their cognitive and emotive sources influenced by a distinctively East Asian cultural philosophy.

Towards a framework: 21st-century East Asian mobilities consumption and Veblen's conspicuous consumption

Given the world's biggest consumers, i.e., a quarter of world's population with Confucian heritage, it is worth giving a thought to how Veblen's conspicuous consumption is met by and enacted by the Confucian heritage consumers. Accordingly, it makes sense to gain a clear understanding of the cultural philosophy, on which the world's biggest consumers' worldviews rest. Confucian five constants are the set of values that gives the fundamental base for the East Asian cultural philosophy of Confucianism. Thus, the focus of aligning Veblen's conspicuous consumption rests on the five Confucian constants. These constants have been witnessed as the guiding lights in East Asian societies for over 2,000 years and have been adopted and applied according to specific time and place conditions. Ren (仁), as part of the five constants, refers to love and benevolence. It highly values tolerance, forgiveness, deference, filial obedience (to parents), faithfulness (to master), wisdom, and honesty. Yi (義 righteousness) places utmost significance on rightness that exists in man's relationship with others. Li (禮 propriety or rites) values the moral force, which governs correct and appropriate behaviour and men's relationship with each other and reflected through personal loyalty. Xin (信 sincerity or trustworthiness) places high esteem on sincerity and trustworthiness of a person, extending to the credibility of his words and actions. Zhi (智 wisdom) appreciates the essential meaning in wisdom and knowledge needed for self-cultivation.

As one of most enduring values of human cultures, Confucian values are expressed and captured accordingly in the form of different actions or cultural behaviours. Chinese tourist motivation research approached from Confucian values uses family, harmony, face, relationship, self-development, and learning as important dimensions to understand Chinese tourist motivations (Fu, Cai, & Lehto, 2017). These dimensions relate to the classic five constants, which were adopted and applied to the relevant time and place of the research. Other research has identified that these fundamental Confucian values can be attributed to East Asian travel behaviours and in particular the ways in which Chinese tourists shop while travelling (Kwek & Lee, 2010; Wang, Bao, Wang, & Wu, 2017). Elsewhere, using an exploratory approach, the Confucian five constants were found to provide an influential philosophical base for hospitality business ethics, leadership, and service provision (Chon & Hao, 2020).

Given the inherent and profoundly influencing nature of Confucian cultural values on East Asian worldviews and everyday life, it is important to consider the potentially related elements of the five constants and Veblen's conspicuous consumption, specifically, conspicuous consumption that overtakes leisure as in his prediction. This would provide us with a more comprehensive knowledge, which could then enable us to see the ways in which East Asian consumers adopt conspicuous consumption and the interactions between the adopted notion (Veblen's conspicuous consumption) and East Asian cultural–philosophical performances of it. Figure 2.2 illustrates conspicuous consumption beyond Veblen's time and space and the five Confucian constants.

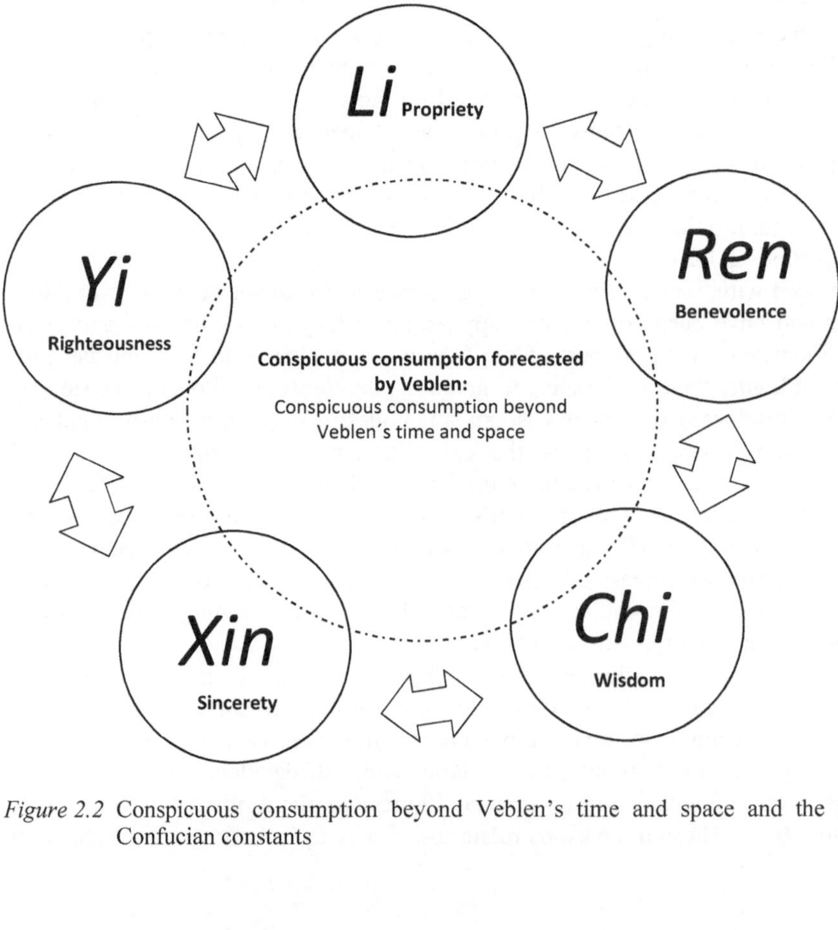

Figure 2.2 Conspicuous consumption beyond Veblen's time and space and the five Confucian constants

There are some reflective points worth recounting regarding the framework, particularly related to leisure and mobilities. Veblen's leisure class and the very notion of leisure are firmly positioned within the linear progressive thought of Western worldview where changes through time in human history mean becoming better or progressing. In this worldview, Veblen points to the essence of leisure being non-productive consumption of time and material. In order to be more efficient in demonstrating one's ability to waste, endowed with excessive accumulation of wealth, one employs conspicuous consumption. While giving rise to the conceptualisation of conspicuous consumption, Veblen's leisure concept does not seem to have moved along with the all-engulfing conspicuous consumption, as forecasted by Veblen. Instead, Veblen's concept of leisure appears to have ceased to exist when leisure was studied as part of social sciences enquiries beyond Veblen's time and space. Indeed, leisure has been defined as:

> Leisure is a period of time referred to as *discretionary time*. It is that period of time when an individual feels no sense of economic, legal, moral or social compulsion or obligation nor of physiological necessity. The choice of how to utilize [*sic*] this time period is solely his. In leisure time, an individual feels he does not "have to" do anything, where *have* refers to the various states of constraint described above.
>
> (Voss, 1967, p. 101 italics in original)

The conceptualisation of leisure in the 20th century not only differs from that of Veblen's but also was refined into two separate domains. Voss continues:

> The remainder of nonwork time is referred to as nondiscretionary time. Here, the individual is faced with a sense of legal, moral, social, or physiological compulsion or obligation when deciding how to allocate his [*sic*] time. He [*sic*] is not, however, faced with a sense of economic obligation. But here, as in work, the individual feels he [*sic*] does "have to" do something.
>
> (Voss, 1967, p. 101)

When the phenomenon of leisure received scholarly attention in the 20th century, its definition of leisure did not coincide with Veblen's leisure conceptually. In fact, 20th-century leisure and its onward investigations of the concept were more occupied with the width and depth of the concept and the universality of it across different cultures in an increasingly globalising world.

Providing a deeper understanding on the concept of leisure, Kelly (1972) proposed a four-celled conceptualisation of leisure where "freedom-discretion" and "work-related" dimensions provided foci. In this conceptualisation, Cell 1 denotes the leisure choice independent of work; Cell 2 refers to the leisure choice related to work; Cell 3 points to leisure activity determined by structural or symbolic social factors although its form and content was independent of work; Cell 4 contains network activity that is socially determined and related to work in form and content (p.55–58). In Kelly's analysis, we can see that Veblen's concept of leisure seems to relate to Cell 3, where people had no liberty of choice for leisure activity as certain social forces and culture already determined in which leisure activities an individual should participate in *per se*. The West-defined concept of leisure has been scrutinised from an East Asian viewpoint. Tracing the etymological origin of a Chinese-relevant leisure concept, Lie, Yeh, Chick, & Zinn (2008) demonstrated that the notion of Chinese leisure could be traced to the country's cultural philosophies of Taoism and Confucianism. It is commendable to highlight the existence of a Chinese concept of leisure. Other ways in which East Asian cultural philosophies further develop through interactions with the West are yet to be researched.

While the conceptualisations of leisure between Veblen and 20th-century researchers seem not to be identical, consumption related to leisure, tourism, and mobilities throughout the 20th and 21st centuries appears to remain constant. Beginning with enquiries into non-work time spheres, the 20th-century conceptualisation of leisure may be argued, at least partly, as a representation of 18th-century industrialisation permeating into structural and institutional settings. With the notion of leisure expanding throughout the globe, facilitated by replicated modernisation and industrialisation beyond the Euro/Western part of the globe, tourism appears to have taken its place in researchers' attention along with leisure. In its continuation of the consumption-focus in leisure and tourism enquiries, recent research argued that tourism consumption needs to be understood as global affecting a local environment (Kirillova, Wang, Fu, & Lehto, 2020). Drawing from a study that included the contexts of China, Russia, and the USA, various society-level and individual-level factors were found to contribute towards consumption forming forces. With leisure and tourism appearing to capture all aspects of human movements in an increasingly connected world. By the early 21st century, social scientists began to recognise a much bigger global structure of movements of humans, machinery, information, and ways of thinking, among others that can be termed as "mobilities". Along with the broadening horizon of leisure and tourism that gave central emphasis on human movements around the globe, e.g. culture and materials, mobilities views human and associated movements as part of larger mobilities (Sheller & Urry, 2004; Winter, 2009; Harrison, 2017). This view extends to the notion of moving consumption itself (Brembeck, Cochoy, & Moisander, 2015). Moving consumption entails the movement of markets, e.g. consumers, goods, political issues such as environmental issues. Given the relationship that the Arctic is forging with Asia, mobilities consumption is expected to have more nuanced and broader interactions between East Asia and a changing Arctic.

Conclusion

Veblen's conspicuous consumption has been adopted to explain Asian consumption patterns and behaviour throughout the 20th and 21st centuries. A critical evaluation of Veblen's much-cited book, *The theory of the leisure class*, identified two types of conspicuous consumption. One type denotes conspicuous consumption employed within the leisure concept defined by Veblen based upon his time and space. The second is the forecast by Veblen that such consumption would absorb all material and social domains, with livelihoods barely being kept from its engulfing. Forecasts for times and spaces beyond Veblen's own epoch, specifically, 21st-century Asian mobilities consumption, may be better understood when we appreciate the dynamic interplays between the second form of conspicuous consumption and the enduring cultural philosophy of the Confucian five constants.

References

Belk, R. W. (1988). Possessions and the extended self. *Journal of Consumer Research, 2*(September), 139–68.

Brembeck, U., Cochoy, F., & Moisander, J. (2015). Moving consumption. *Consumption Markets & Culture, 18*(1), 1–9. doi:10.1080/10253866.2014.899218

Chon, K., & Hao, F. (2020). The five constants: A Confucian business model for the hospitality industry. *Journal of Hospitality & Tourism Research, 44*(7), 1051–1071. doi:10.1177/1096348020936342.

Cordes, C. (2005). Veblen's "Instinct of workmanship," its cognitive foundations, and some implications for economic theory. *Journal of Economic Issues, 39*(1),1–20. Retrieved from http://www.jstor.com/stable/4228111

Fu, X., Cai, L., & Lehto, X. (2017). Framing Chinese tourist motivations through the lenses of Confucianism. *Journal of Travel & Tourism Marketing, 34*(2), 149–170.

Harrison, D. (2017). Tourists, mobilities and paradigms. *Tourism Management, 63*, 329–337. doi:10.1016/j.tourman.2017.07.002

Hodgson, G. M. (2008). How Veblen generalized Darwinism. *Journal of Economic Issues, 42*(2), 399–405. doi:10.1080/00213624.2008.11507148.

Huimei, L., Yeh, C. K., Chick, G. E., & Zinn, H. C. (2008). An exploration of meanings of leisure: A Chinese perspective. *Leisure Sciences, 30*(5), 482–488. doi:10.1080/01490400802353257

Kelly, J. R. (1972). Work and leisure: A simplified paradigm. *Journal of Leisure Research, 41*, Winter, 439–451.

Kirillova, K. D., Wang, X., Fu, X., & Lehto, X. (2020). Beyond "culture": A comparative study of forces structuring tourism consumption. *Annals of Tourism Research, 83*, 102941. doi:10.1016/j.annals.2020.102941

Kwek, A., & Lee, Y. S. (2010). Chinese tourists and Confucianism. *Asia Pacific Journal of Tourism Research, 15*(2), 129–141.

Li, J., Guo, S., Zhang, J. Z., & Sun, L. (2019). When others show off my brand: Self-brand association and conspicuous consumption. *Asia Pacific Journal of Marketing and Logistics, 32*(6),1214–1225. doi:10.1108/APJML-04-2019-0225

McCormick, K. (2011). Thorstein Veblen. In G. Ritzer & J. Stepnisky (Eds.), *The Wiley-Blackwell companion to major social theorists, Vol. 1, Classical social theorists* (pp. 185–204). Malden: Wiley-Blackwell.

Park, I., & Cho, L. J. (1995). Confucianism and the Korean family. *Journal of Comparative Family Studies 26*(1), 117.

Park, K.-S., & Reisinger, Y. (2009). Cultural differences in shopping for luxury goods: Western, Asian, and Hispanic tourists. *Journal of Travel & Tourism Marketing, 26*(8), 762–777. doi:10.1080/10548400903356152

Phau, I., & Teah, M. (2009). Devil wears (counterfeit) Prada: A study of antecedents and outcomes of attitudes towards counterfeits of luxury brands. *Journal of Consumer Marketing, 26*(1),15–27. doi: 10.1108/07363760910927019

Plotkin, S. (2017). Introduction: Thorstein Veblen's elusive project. In S. Plotkin (Ed.), *Anthem companion to Thorstein Veblen* (pp. 1–18). Anthem Companions to Sociology. London: Anthem Press.

Rogers, J. A. (1972). Darwinism and social Darwinism. *Journal of the History of Ideas, 33*(2), 265–280. Retrieved from http://www.jstor.com/stable/2708873

Sheller, M., & Urry, J. (Eds.). (2004). *Tourism mobilities: Places to play*. London: Routledge.

Taylor, D. G., & Strutton, D. (2016). Does Facebook usage lead to conspicuous consumption?: The role of envy, narcissism and self-promotion. *Journal of Research in Interactive Marketing, 10*, 231–248.

UNWTO. (2018). Tourism highlights 2018 edition. Retrieved September 11, 2020, from https://www.e-unwto.org/doi/pdf/10.18111/9789284419876

Veblen, T. (1899/2009). *The theory of the leisure class*. New York: Oxford University Press.

Voss, J. (1967). The definition of leisure. *Journal of Economic Issues, 1*, 91–106.

Waller, W. (2017). Reconsidering Thorstein Veblen's use of instincts. In S. Plotkin (Ed.), *Anthem companion to Thorstein Veblen* (pp. 39–67). Anthem Companions to Sociology. London: Anthem Press.

Wang, J., Bao, J., Wang, C., & Wu, L. (2017). The impact of different emotional appeals on the purchase intention for green products: The moderating effects of green involvement and Confucian cultures. *Sustainable Cities and Society, 34*, 32–42.

Winter, T. (2009). Asian tourism and the retreat of Anglo-western centrism in tourism theory. *Current Issues in Tourism, 12*(1), 21–31. doi:10.1080/13683500802220695

Zakaria, N., Wan-Ismail, W.-N. A., & Abdul-Talib, A-N. (2020). Seriously, conspicuous consumption? The impact of culture, materialism and religiosity on Malaysian Generation Y consumers' purchasing of foreign brands. *Asia Pacific Journal of Marketing and Logistics*, Ahead-of-print. doi:10.1108/APJML-07-2018-0283

Part 2

Arctic politics and Asian mobilities

3 Chinese tourism consumption vis-à-vis tourism development strategies in the Arctic

Emma Björner, Elena Dybtsyna, and Erik Lundberg

Introduction

Asian tourism consumption in Arctic destinations can be explained as a reflection of the global impetus for the Arctic (Lee, 2020). The Arctic represents an interesting context as a region of rich natural resources, such as fish, oil, gas, minerals, tourism, and transport solutions by sea. The Arctic is also of strategic importance for sustainable development. Improved access to Arctic resources, new employment opportunities, and an emerging influence from international actors outside the region imply challenges and opportunities for the sustainable development of the Arctic (Heininen, Everett, Padrtová, & Reissell, 2020).

Further, the Arctic is a cross-border area with its own purposes and process dynamics. Arctic policies are developed on a national level (national Arctic policies), intergovernmental level (Arctic Council, Barents Secretariat), and international level (European Union, EU, Arctic Policy). We are interested in policies and strategies because they provide clues as to why and how tourism development is regarded as important for sustainable development in the Arctic. The Nordic Arctic—consisting of the counties Norrbotten and Västerbotten in Sweden (also known as Swedish Lapland); Troms and Finnmark, and Nordland in Norway; and the regions of Lapland, Northern Ostrobothnia and Kainuu in Finland—share common economic, environmental, and social interests. Moreover, the Expert Group report set up by the Prime Ministers of Norway, Sweden, and Finland (Prime Minister's Office, 2015) concludes that tourism has become a major source of income in northern Sweden, Norway, and Finland and has been successful in attracting tourists from all over the world.

The current national Arctic strategies of states such as Sweden, Norway, and Finland consider tourism as a key industry in the Arctic and point out its significance for sustainable development in the region. Moreover, China's Arctic Policy centres on China being the source of tourists to the Arctic. The tourism industry's own strategies also recognise that Asian markets show a very strong growth in the number of travellers over the last decade and that there is great potential in the longer term, even if this growth was halted in 2020, due to the COVID-19 crisis. Before this disruptive event, Chinese outbound tourism had become the largest outbound market, outgrowing traditional strongholds such as the US and

DOI: 10.4324/9781003039518-3

German markets (Lee, 2020). It is the largest outbound market both in departure numbers and in terms of total expenditure and is characterised by its large volume, rapid growth, and vast potential (United Nations World Tourism Organisation, UNWTO, 2019). This growth has been rapid, since the Chinese outbound market did practically not exist before the 1990s when travelling was very restricted (Lee, 2020). Various factors have contributed to the sustained strength of the Chinese outbound tourism market, such as higher incomes and more overseas experience, as well as access to online travel information, air connectivity, and accessibility. Consequently, the Arctic strategies in the Nordic countries would have to consider this reality.

China has no territorial sovereignty in the Arctic but has developed a self-defined Arctic identity as a "near-Arctic state". In 2013, China moreover gained observer status in the Arctic Council (Grieger, 2018), and in 2018, China published its first white paper on Arctic policy, acknowledging that its Arctic interests extend to a variety of commercial activities. It is moreover recognised that China is among the international actors outside of the Arctic region with emerging influence in setting future agendas on Arctic development (Lim, 2018). Furthermore, China identifies "Arctic tourism" as "an emerging industry" with China being a source of tourists to the Arctic (China's Arctic Policy, 2018).

The present chapter addresses the following research questions: (1) how is tourism development and destination branding portrayed in policy and strategy documents dealing with the Arctic and (2) how does it correspond with Chinese tourism consumption in the case of the Swedish Arctic? As such, this chapter considers current Arctic national policies and strategies of Sweden, Norway, and Finland, and especially Sweden's, as well as China's. For the purpose of this book, we have decided to focus on the Nordic Arctic and in particular Sweden as an example of an Arctic state (in addition to chapters dealing with other Arctic states) and to contribute to the discussion by considering the case of the Swedish Arctic when it comes to tourism development and Chinese tourism consumption.

Tourism policy and strategy

The analysis of strategic planning and policy documents is a prolific tourism research area used to discuss tourism development. The rationale for such analysis is that these documents express how governments (local, regional, and national) view development and how it should be achieved and that the wording often has a direct influence on actual development and industry decisions (Connell, Page, & Bentley, 2009; Moyle, McLennan, Ruhanen, & Weiler, 2014). Dredge and Jamal (2015) remind us that policy is not created in a vacuum and that the context in which it is created is becoming increasingly important (with the rise of corporate influence, public-private partnerships, non-government organisations, etc.).

One reason for the importance of policy and strategy documents in tourism is the fragmented nature of the tourism industry, which is constituted of numerous small and medium-sized enterprises in various sectors. Policy can create cohesion, coordination, and integration, in particular on local government levels

(Dredge, 2006; Dredge, Jenkins, & Whitford, 2011). The documents analysed in this chapter are viewed through these lenses. The wordings of policies and strategies are considered to determine if they influence tourism activities and Chinese tourism consumption, and, if there are any signs of cohesion, coordination, and integration.

Since policies and strategies are not created in a vacuum, it is useful to analyse how and why policies evolve. Hall (2011) uses the concepts of policy failure and learning in the context of sustainable tourism governance. Of particular interest in the context of this chapter is his notion of first-, second-, and third-order change, i.e. how policies are evolving or changing in relation to internal and external factors. First-order change implies internally driven, small-scale, incremental policy change. Second-order change is also, largely, internal work among officials and policy experts. However, realising that policy goals are not met, new policy instruments and techniques are introduced. In the context of tourism in the Arctic, it could, for example, mean new techniques for measuring sustainable tourism growth. However, it does not mean a radical change of policy goals. The latter is fundamental for third-order change, or policy paradigm shifts, where a new set of policy goals are formulated based on external influence and the realisation that current goals are failing or are incoherent (Hall, 2011). In an Arctic context, it could, for example, imply a shift from a focus on economic growth to environmental conservation in terms of tourism development. The process of policy paradigm shifts could help understand why and how policies and strategies evolve.

Another useful concept, from dependency theory is path dependency (cf. Gill & Williams, 2011; Grabher, 1993). It shows how historical decisions, in, for example, a destination, influence the current development even if they might be irrational or irrelevant. A chosen path creates "lock-ins", where it is difficult to see other possibilities for development. When prior decisions have led to a particular development path, previous research shows that destinations have difficulties in changing direction (Gill & Williams, 2011; Lundberg, 2015) from, for example, unsustainable to sustainable tourism development. As such, path dependency and "lock-ins" can be useful in order to understand governance and strategies in tourism development in the Swedish Arctic and in particular discrepancies between policy and strategy for industry and regions (e.g. national and regional levels) or between countries (e.g. Sweden and Finland) (Gill & Williams, 2011).

Tourism consumption and Confucian values

To iterate, we aimed to understand how tourism development and destination branding, as portrayed in strategies and policies, correspond with Chinese tourism consumption. For lenses, we have also used the concepts of conspicuous consumption and the Confucian Five Constants, introduced in Chapter 2. Conspicuous consumption has repeatedly been applied to understand tourism consumption in general (e.g. Phillips & Back, 2011) and Chinese tourism consumption in particular (e.g. Hung, 2018). Conspicuous consumption originates from Veblen's theory of the leisure class, a framework in which preferences are determined socially in relation to the positions of individuals in the social hierarchy. Veblen also claimed that individuals imitate the consumption patterns of other individuals situated at higher points in the hierarchy. The social norms that govern such imitation change as the economy and its social fabric evolve over time (Trigg, 2001).

Veblen's theory has, however, been criticised (in this book and elsewhere). It has been argued that Veblen's approach is too limited in relying on the "trickle down" of consumption patterns from the top of the social hierarchy (Trigg, 2011). Pacesetters for consumption may also be those at the bottom of the social hierarchy (Fine & Leopold, 1993; Lears, 1993). Furthermore, consumers no longer always want to display their wealth conspicuously but convey status in more subtle and sophisticated ways (Canterbery, 1998; Mason, 1998). Moreover, consumer behaviour is no longer formed by positions of social class but by lifestyles that cut across social hierarchies (Featherstone, 1991; McIntyre, 1992). Relatedly, conspicuous consumption that expresses a successful and luxurious life is no longer the domain of so-called *A-listers* but also "ordinary" consumers (Rokka & Canniford, 2016). In a Chinese context, consumers engage in conspicuous consumption for various reasons: some to disassociate from the crowd, others to associate with the aspiration group (Huang & Wang, 2018).

Description of secondary data and analysis

In order to answer the research questions proposed in this chapter, a number of policy and strategy documents have been reviewed and analysed. These include current Arctic national strategies for Sweden (Government Offices of Sweden, 2011, 2020), Norway (Norwegian Ministries, 2017), Finland (Prime Minister's Office, Finland, 2013), and China (People's Republic of China, The State Council, 2018); Arctic business development reports (Prime Minister's Office, 2015; *Sustainable business development in the Nordic Arctic* by Olsen et al., 2016); tourism policies and strategies for Sweden, Norway, and Finland (Nordic Tourism Policy Analysis (Nordic Council of Ministers, 2019); Plan for Nordic Cooperation 2019–2023, (Nordic Council of Ministers, 2019), and for China (National Tourism and Leisure, 2013 [General Office and the State Council, People's Republic of China, 2013]); as well as national and regional tourism policies and strategies centring on Sweden (National Strategy for Swedish Tourism 2020 (Svensk Turism AB, 2010); About Swedish Lapland (Swedish Lapland Visitors Board, 2020). In addition, material consisting of reports, literature, and media coverage about Chinese tourism consumption overseas were also reviewed and analysed. Some material was also reviewed in order to gain a holistic understanding of Arctic tourism development. These included, for example, reports produced by the Prime Minister Office of Finland and Nordregio (Nordic Centre for Spatial Development, under the auspices of the Nordic Council of Ministers), which deal with sustainable business development of the Nordic Arctic in Sweden, Norway, and Finland. Figure 3.1 offers a visualisation of the material reviewed and analysed.

The authors of this chapter conducted the analysis of the material just described and performed a re-analysis of data collected by others (Hakim, 2000). The

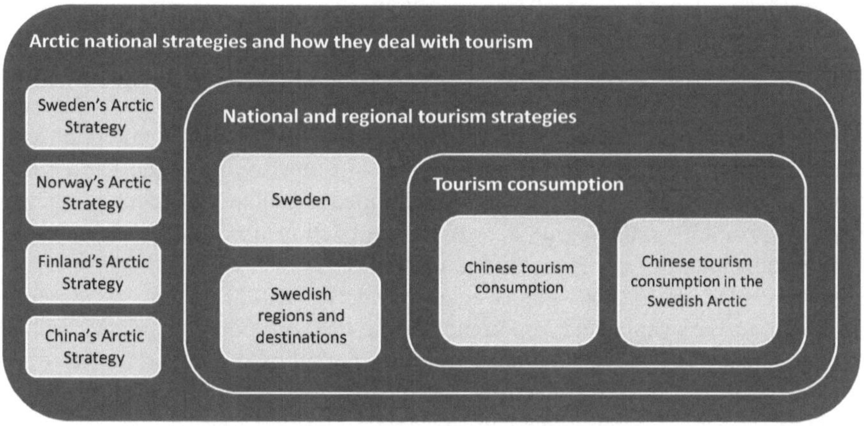

Figure 3.1 Visualisation of the material reviewed and analysed

authors read the material with the same "frame of reference" in the sense that we all paid attention to the tourism industry as well as the research questions and the scope of this chapter. Previous research and literature on tourism policy and strategies, tourism development, and destination branding, as well as conspicuous consumption and the Confucian Five Constants, as previously mentioned, were used as lenses to interpret the material and perform the analysis found in the subsequent sections.

Findings and discussion

Arctic nations' and China's policies and strategies and how they deal with tourism

The Arctic national policies of Sweden, Norway, and Finland pursue similar priorities when it comes to the sustainable economic development of their Arctic regions, including priorities related to tourism. Nature-based experiences, the opportunity to see the midnight sun, and the northern lights are attractions shared by all destinations (Karlsdottir, Olsen, Harbo, Jungsberg, & Rasmussen, 2017). Arctic tourism has proven attractive for tourists as well as a lucrative industry that brings thousands of visitors to the Nordic Arctic. Maintaining balanced development in terms of economic, socio-cultural, and environmentally sustainable development is challenging, which is why it needs to be addressed in the development of regional and local strategies for tourism development (Olsen et al., 2016).

Sweden's *Strategy for the Arctic Region* (Government Offices of Sweden, 2011) emphasises that the tourism industry in the Arctic has growth potential and recognises that the Arctic, with its nature and access to clear air, water, and silence, is in demand by many visitors. It also highlights that the tourism industry has a significant potential for job creation and boosting sustainable economic growth. This is almost in unison and repeated in Sweden's *Strategy for the Arctic Region* (Government Offices of Sweden, 2020). The strategy presented in 2020 states that Sweden intends to cooperate with different actors and recognise the region's sensitive environment and the needs and situation of its indigenous people to develop sustainable and attractive tourist destinations. It also emphasises (in contrast with the strategy from 2011) that Sweden's growth potential and Sweden's Arctic profile is in both traditional and new tourism industries, such as culture and gastronomy. However, no additional financial resources to realise Sweden's priorities in the region nor government actions for the tourism industry in the Swedish Arctic are present in the strategies.

Norway's Arctic Strategy (Norwegian Ministries, 2017) recognises the current and future relevance and importance of the tourism industry, and tourism is increasing, with visitors coming to the region from all over the world. The strategy further states that it will ensure economically, socially, and environmentally sustainable business development in the Arctic and that tourism has considerable potential for the future. However, the strategy also points out that the policy choices towards development in northern Norway are to be designed to facilitate increased activity within an environmentally sound framework. In addition, the suggestion for the government's budget for 2021 is earmarked at NOK 255 million for initiatives in northern Norway to promote Norwegian interests and to contribute to realising the government's priorities in the region (Holm, 2020).

Finland's Strategy for the Arctic Region (Prime Minister's Office, Finland, 2013) identifies the tourism sector as one of its key priorities. It also aims for competitiveness and sustainable growth. Finland's objective is the increase and renewal of the tourism sector in accordance with the three pillars of sustainability. Finland's strategy moreover emphasises the necessity of taking care of natural scenery, environment, and the uniqueness of the local culture. The strategy also recognises that tourism jobs cannot be relocated. Additionally, Finland's strategy comes up with a detailed action plan to be undertaken by the tourism sector in the Finnish Arctic. These actions include recognition of Arctic tourism as a growth sector and government support for the region in terms of land use planning, investments, and sufficient development resources.

China's Arctic Policy (People's Republic of China, The State Council, 2018) outlines China's policy goals, centring on understanding, protecting, developing, and participating in the governance of the Arctic. The Chinese strategy recognises that Arctic tourism is an emerging industry. China also wants to contribute to the sustainable development of the Arctic by raising the environmental awareness of Chinese tourists as well as advocating low-carbon tourism, ecotourism, and responsible tourism. The strategy also states that China is a source of tourists to the Arctic, and China supports and encourages its enterprises to cooperate with Arctic states in developing tourism in the region.

The findings indicate that Sweden, Norway, and Finland, as well as China, increasingly focus on the economic, environmental, and social sustainability of the tourism industry in the Arctic. In addition, we see that Sweden's new Arctic Strategy (Government Offices of Sweden, 2020) encourages Arctic profiles for tourist destinations. What is different between the countries studied is that Sweden, in both Arctic strategies, lacks national initiatives aimed at tourism development in terms of funding and government actions. As such, Sweden appears to be "locked-in"(cf. Gill & Williams, 2011) when it comes to not offering funding and government support to the Swedish Arctic region.

National and regional level tourism policies and strategies in Sweden

The National Strategy for Swedish Tourism was launched in 2010. A central vision was to double the turnover in the tourism sector from 2010 to 2020, and the way to do that was by focusing on coordination, proactive marketing, and development of Swedish destinations. A central aim was also to contribute to the positive development for the Swedish tourism sector, with a focus on stable and long-term sustainable growth, increased national and international competitiveness, and increased market shares for Swedish tourism, while also preserving natural and cultural values.

In Sweden's national strategy, tourism development and destination branding are discussed in relation to developing world class destinations, regional development, and viability, as well as development of the tourism industry and tourism products. The branding of Swedish destinations and Sweden as a nation, and the development of Sweden's attractiveness and the image of Sweden, is also emphasised. The strategy points out that the Swedish tourism industry has the opportunity to take advantage of the unique image of Sweden. This can be achieved by communicating Sweden as a progressive and sustainable travel destination, which is based on Sweden's unique characteristics founded upon four core values, namely innovation, openness, authenticity, and caring. These core values can be interpreted in relation to the Five Constants. Caring can be seen as tied to *Ren*, and thus kindness, compassion, generosity, and respect (Chon & Hao, 2020), which, for example, could be manifested in the hospitality of Swedish tourism actors. Authenticity can be seen as related to *Xin,* and thus trustworthiness and credibility (ibid), which can be tied to both a match between the communicated image of Swedish destinations and the reality experienced as well as experiencing the "real" Sweden.

Unique experiences, attractions, assets, values, traits and the unique image of Sweden is elaborated upon in the strategy. Sweden's unique assets are exemplified with nature, winter, the environment, Swedish cities, restaurants, culture, creativity, design, and cultural values. Unique experiences can be related both to the notion of conspicuous consumption and to the Confucian constant *Zhi* and thus wisdom and knowledge as well as a willingness to explore and learn from experience (Chon & Hao, 2020).

When it comes to regional level tourism policies and strategies in Sweden, we focus on Swedish Lapland, located in the counties of Norrbotten and Västerbotten, which are described as Sweden's part of the Arctic. Swedish Lapland adopted the same goal as Sweden's national tourism strategy, namely to double the turnover (Tourism in Swedish Lapland, 2020). Swedish Lapland has also set the goal that more tourism actors become professional businesses that can receive foreign tourists (Olsen et al., 2016).

One of the strategic priorities of Swedish Lapland Visitors Board (2020) is to work more consciously towards sustainable development and to consider not only economic sustainability but also social and environmental sustainability in all development initiatives. Key themes in the materials include destination development, destination branding, sustainability, uniqueness, experiences, nature, and the Arctic. In terms of the development of destinations, emphasis is on joint efforts to maximise the opportunities for growth and to ensure the long-term and sustainable development of the region. Swedish Lapland is also depicted as a common brand and cooperation platform for local destinations and tourism businesses in the region, with the aim to be competitive, to brand north Sweden as a destination, and to attract international visitors (Swedish Lapland Visitors Board, 2020).

The destination Swedish Lapland is described as holding a unique position in global tourism through its subarctic nature and culture and influenced by the indigenous Sámi people living in the region. It is also depicted as an internationally attractive destination with international guests and especially affluent ones, and luxury tourism is set to increase in the region. The Ice Hotel, Tree Hotel, and Arctic Bath are put forth as exclusive and unique experiences. This focus on luxury tourism and unique experiences can be tied to conspicuous consumption and the Five Constants, especially *Zhi* and thus experience-based wisdom (Chon & Hao, 2020).

The findings show that there appears to be alignment between policies and strategies regarding the Swedish Arctic at national and regional levels. As previously intimated, path dependency (cf. Gill & Williams, 2011; Grabher, 1993) can assist in explaining the creation of "lock-ins" where it is difficult to see other possibilities for development or changed direction (Gill & Williams, 2011; Lundberg, 2015), for example, when it comes to sustainable tourism development. In the case of Swedish policies and strategies, a history of focusing on growth, nature, conservation, and the environment seem to permeate and influence destination development. These foci reveal a possible conflict between growth and environmental sustainability, in particular when the policy goal has been to double the turnover of tourism. This "lock-in" could hinder much needed second- and third-order changes (Hall, 2011).

Our findings show that the Swedish Government and local authorities consider the tourism industry in the Arctic as having potential and a strategic impact and that it encourages the private sector to invest in the Arctic region. These efforts are also in line with Finnish and Norwegian policies and strategies that seek to attain the same priorities when it comes to the economic development outlined in their Arctic strategies. However, what is lacking in Sweden's Arctic strategies are the action plans and resources that can be allocated to sustainable tourism development of the Swedish Arctic region. As Khorrami (2019) argues, the unique nature, creative culture, and sustainably produced food should be turned into potential sources of economic growth as part of tourism development and destination branding. These are efforts that also are emphasised in the Swedish strategies.

While Sweden's Arctic and regional strategies take the Arctic context into consideration, Sweden's national strategy is more generally formulated and lacks a specific focus on Arctic destinations. Furthermore, national and regional tourism strategies in Sweden set goals for Arctic destinations to be competitive in a sustainable way, centring on all aspects of sustainability yet still with more emphasis on environmental and economic sustainability. Consequently, it appears as if there is a need for a third-order change or a policy paradigm shift (Hall, 2011) where an updated set of policy goals—that align with and put emphasis on all three pillars of sustainability—are formulated in the Swedish Arctic strategy, Sweden's national strategy, and in strategies on regional levels.

The findings also show that Chinese tourism consumption in the Arctic is recognised by China's Arctic strategy, which considers Chinese tourism as the main source of tourists to the Arctic. For the Nordic Arctic, it is vital to understand if this new, growing market segment is different from the more traditional markets, in particular, to adopt policies and strategies to cater to their needs and to avoid potential negative tourism impacts, which are elaborated on next.

Arctic and Swedish policies and strategies vis-à-vis Chinese tourism consumption

Like many other tourists, the Chinese are interested in experiences that can offer natural scenery, high value for money, family-friendly locales, and experiences that can contribute to well-being, self-realisation, and personal development (Huang & Wang, 2018). Furthermore, Chinese tourists to the Arctic are not generally the typical mass tourist but more independent travellers, similar to the experienced Western traveller (Cai, Ma, & Lee, 2020). This is important information for policy makers and local entrepreneurs in the Swedish Arctic to take into account. Thus, it appears relevant that national and regional tourism strategies do not focus on Chinese tourists specifically but consider them as a type of experienced traveller. However, value orientations such as self-cultivation (cf. Pearce & Lee, 2005), collectivism, and human-environment harmony (Cai et al., 2020) of Chinese tourists influence their behaviour and motivations. In particular, the latter could influence behaviour in a Swedish Arctic context where nature is the core attraction (cf. Olsen et al., 2016).

Self-cultivation and harmony can be tied to Confucianism and the Five Constants. "Confucian tourists" are said to achieve their ideal way of life through inner and exterior cultivations and achieve harmony through "guanxi", conformity, deference, and authority (Chon & Hao, 2020). Self-cultivation is a vital moral code in the Confucian Constant *Yi* through its focus on moral judgement and making the right actions based on good judgement. Human-environment harmony can furthermore be related to the Confucian Constant *Ren* by its focus on humanness, compassion, and respect as well as human empathic reciprocity and moral sentiments (ibid).

Consequently, there are indications that Chinese tourists have certain traits that need to be taken into account. One example is the significant cultural value of "saving face" which includes activities such as gift purchasing and giving and is related to social status and relationship building (Kwek & Lee, 2015). The cultural value of saving face may consequently differentiate Chinese tourists from other tourists and impact Chinese tourists' tourism consumption. Zhang, Pearce, and Chen (2019) moreover address what is often in the media discussed as Chinese tourists' "uncivilised" behaviour through the lenses of "face" and social identity (cf. Kwek & Lee, 2015). They find that, as Chinese tourists gain tourism experience, they want to accommodate for potential cultural conflicts with hosts in order to preserve both their own and the collective's face. Gaining tourism experience ties into the Confucian constant *Zhi*, meaning both wisdom (智) and thus the ability to learn, as well as knowledge (知) and thus the result of learning. Zhi has also been conceptualised as experience-based and morality-based wisdom (Chon & Hao, 2020). Hosts in the Swedish Arctic can facilitate Chinese tourists' ambitions and efforts to gain tourism experience by, for example, informing Chinese tourists of local traditions, customs, and behaviours (Zhang et al., 2019), which relates to the Confucian Constant *Li*, meaning rituals, rites, etiquette, courtesy, and politeness (Chon & Hao, 2020).

Furthermore, a number of outdated impressions and myths remain in the tourism industry when it comes to Chinese tourists (Dichter, Chen, Saxon, Yu & Suo, 2018). Contrary to many myths, Chinese tourists value experience over shopping, and are not only interested in iconic landmarks but have diverse travel needs. They often tend to stay longer in one place, are still interested in (especially high-end, semi self-guided) package tours, value fine dining and various cuisines, and are used to e-payment through Alipay and WeChat. Moreover, the beliefs of family, friends, and peers play a central role when making decisions about destinations, and Chinese travellers are inclined to use various digital channels to get information and make bookings for accommodation and activities (Dichter et al., 2018). In order to meet the future demand of Chinese tourists to the Swedish Arctic, there is a need for policies and strategies to take into account the traits of Chinese consumers in relation to tourism development and destination branding.

In international tourism, a trend points to an increase in the market for experiences and destinations that can contribute to well-being, self-realisation, and personal development. Tourism consumption is predicted to take on new expressions with increased prioritisation for experiences in contrast to goods and services (Invest in Norrbotten, 2020). Experience is also something that matters more and more for Chinese tourists who venture further afield and desire places that offer natural scenery, high value for money, and family-friendly locales (Dichter et al., 2018). Seen through the lens of the Confucian constant *Zhi* (wisdom and knowledge), a focus on experiences when travelling can be seen as attractive to Chinese tourists as it offers possibilities for experience-based learning (Chon & Hao, 2020). Furthermore, Chinese tourists' interest in engaging in unique and exclusive tourism experiences can be tied to conspicuous consumption, and thus attempts to disassociate from the crowd or associate with the aspiration group (Huang & Wang, 2018). The Swedish Arctic already offers unique and luxurious experiences that can be attractive for this type of tourist and continues to develop more experiences for the more affluent segment. It can also be worth taking into account that consumers do not always like to display their wealth conspicuously but convey status in more subtle and sophisticated ways (Canterbery, 1998).

Conclusions and further research

We have written this chapter during the global pandemic caused by COVID-19. Tourism was the first industry to be shut down by the impacts of COVID-19, and it will take time to recover. The Arctic Council (2020) states that as the direct economic effects have been significant, indirect impacts depend on how well the tourist sector is integrated into and supported by regional and local levels. Due to the pandemic outbreak, the Arctic region faces unique challenges and risks. When it comes to tourism and hospitality, the most devastating impact may be in the cruise industry and gastronomic industries. They will most probably need several years to return to 2019 levels (Arctic Council, 2020). At the same time, the Arctic has its strengths and resilience, as it is seen as a clean, safe, and sparsely populated region, which tourists may value even more in the future (ibid).

In terms of managerial implications, we would like to emphasise that Swedish national authorities need to achieve an appropriate balance between the three pillars of sustainable development and provide sufficient resources, as in northern Norway, or to come up with government actions, as in northern Finland, to ensure sustainable tourism development in the Swedish Arctic. By doing that, Sweden will be able to support the bottom-up "arctification" (Eklund, 2019) of local businesses post COVID-19 and be able to face future Chinese tourism consumption in Swedish Arctic tourism destinations.

In terms of further research, it would be valuable and relevant to explore Chinese tourism "on the ground" in the Swedish and Nordic Arctic. Conceptually, it would, for example, be fruitful to investigate the social-psychological perspective related to the reversed host-guest relationship stemming from the change from west-east mobility towards more east-west movements (Huang, Keating, Kriz, & Heung, 2015).

References

Arctic Council. (2020). Covid-19 in the Arctic: Briefing document for Senior Arctic Officials. Retrieved from https://oaarchive.arctic-council.org/bitstream/handle/11374/2473/COVID-19-in-the-Arctic-Briefing-to-SAOs_For-Public-Release.pdf?sequence=3&isAllowed=y

Cai, Y., Ma, J., & Lee, Y.-S. (2020). How do Chinese travelers experience the Arctic? *Scandinavian Journal of Hospitality and Tourism*, *20*(2), 144–165.

Canterbery, E. R. (1998). The theory of leisure class and the theory of demand. In W. J. Samuels (Ed.), *The founding of institutional economics* (pp. 139–156). London: Routledge.

Chon, K., & Hao, F. (2020). The five constants: A Confucian business model for the hospitality industry. *Journal of Hospitality & Tourism Research*, 44(7), 1051–1071.

Connell, J., Page, S. J., & Bentley, T. (2009). Towards sustainable tourism planning in New Zealand: Monitoring local government planning under the Resource Management Act. *Tourism Management, 30*(6), 867–877.

Dichter, A., Chen, G., Saxon. S, Yu, J., & Suo, P. (2018). *Chinese tourists: Dispelling the myths—An in-depth look at China's outbound tourist market*. McKinsey & Company. Retrieved from: https://www.mckinsey.com/~/media/mckinsey/industries/travel%20transport%20and%20logistics/our%20insights/huanying%20to%20the%20new%20chinese%20traveler/chinese-tourists-dispelling-the-myths.ashx

Dredge, D. (2006). Policy networks and the local organisation of tourism. *Tourism Management, 27*(1), 269–280.

Dredge, D., & Jamal, T. (2015). Progress in tourism planning and policy: A post-structural perspective on knowledge production. *Tourism Management, 51*, 285–297.

Dredge, D., Jenkins, J., & Whitford, M. (2011). Tourism policy and planning: Historical development and contemporary challenges. In Dredge, D., & Jenkins, J. (Eds.), *Stories of practice: Tourism planning and policy* (pp. 13–34). Aldershot: Ashgate Publishing.

Eklund, N. (2019). The Swedish chairmanship: Foresight and hindsight in Arctic activism. In Nord, D. (Ed.), *Leadership for the North* (pp. 71–88). Springer Polar Sciences. Cham: Springer.

Featherstone, M. (1991). *Consumer culture and postmodernism*. London: Sage.

Fine, B., & Leopold, E. (1993). *The world of consumption*. London: Routledge.

General Office and the State Council, People's Republic of China. (2013). *National tourism and leisure*. Guobanfa, No. 10. Beijing: General Office and the State Council, People's Republic of China.

Gill, A. M., & Williams, P. W. (2011). Rethinking resort growth: Understanding evolving governance strategies in Whistler, British Columbia. *Journal of Sustainable Tourism, 19*(4–5), 629–648.

Government Offices of Sweden. (2011). *Sweden's strategy for the Arctic region*. Stockholm: Regeringskansliet, Government Offices of Sweden.

Government Offices of Sweden. (2020). *Sweden's strategy for the Arctic region*. Stockholm: Government Offices of Sweden.

Grabher, G. (1993). *The embedded firm*. London: Routledge.

Grieger, G. (2018). China's Arctic policy: How China aligns rights and interests. Briefing. European Parliamentary Research Service, PE 620.231, May 2018, European Parliament.

Hakim, C. (2000). *Research design: Successful designs for social and economic research*. Abingdon: Routledge.

Hall, C. M. (2011). Policy learning and policy failure in sustainable tourism governance: From first and second to third order change? *Journal of Sustainable Tourism, 19*, 649–671.

Heininen, L., Everett, K., Padrtová, B., & Reissell, A. (2020). *Arctic policies and strategies — Analysis, synthesis, and trends*. Laxenburg: International Institute for Applied Systems Analysis.

Holm, A. (2020). Statsbudsjettet: Lettere å finne det som avlyses enn det som settes i gang, *High North News*, 7 October 2020.

Huang, S., Keating, B. W., Kriz, A., & Heung, V. (2015). Chinese outbound tourism: An epilogue. *Journal of Travel & Tourism Marketing, 32*(1–2), 153–159.

Huang, Z., & Wang, C. L. (2018). Conspicuous consumption in emerging market: The case of Chinese migrant workers. *Journal of Business Research, 86*, 366–373.

Hung, K. (2018). Consumption of luxury hotel experience in contemporary China: Causality model for conspicuous consumption. *Tourism Review International*, *22*(3–4), 171–185.

Invest in Norrbotten. (2020). *Tourism in Swedish lapland.* Retrieved from https://www.investinnorrbotten.se/tourism-industry/tourism-in-swedish-lapland/

Karlsdottir, A., Olsen, L. S., Harbo, G. H., Jungsberg, L., & Rasmussen, R. O. (2017). *Future regional development policy for the Nordic Arctic: Foresight analysis 2013–2016.* Nordregio Report 2017:1. Stockholm, Sweden: Nordregio.

Khorrami, N. (2019). *Reviewing Stockholm's strategy for sustainable regional development in the Arctic.* Washington: The Arctic Institute. Retrieved from https://www.thearcticinstitute.org/stockholm-strategy-sustainable-regional-development-arctic

Kwek, A., & Lee, Y.-S. (2015). How "face" matters: Chinese corporate tourists in Australia. *Journal of Travel & Tourism Marketing*, *32*(1–2), 120–140.

Lears, T. J. (1993). Beyond Veblen: Remapping consumer culture in twentieth century America. In S. C. Hollander & K. M. Russel (Eds.), *Marketing, Volume II* (pp. 27–40). Aldershot: Edward Elgar.

Lee, Y.-S. (2020). Asia and Arctic tourism. *Scandinavian Journal of Hospitality and Tourism*, 20(2), 105–109.

Lim, K.S. (2018). China's Arctic policy & the polar silk road vision. In Heininen, L., & Exner-Pirot, H. (Eds.), *Arctic yearbook 2018* (pp. 1–17). Akureyri Iceland: Arctic Portal.

Lundberg, E. (2015). The level of tourism development and resident attitudes: A comparative case study of coastal destinations. *Scandinavian Journal of Hospitality and Tourism*, *15*(3), 266–294.

Mason, R. (1998). *The economics of conspicuous consumption: Theory and thoughts since 1700.* Aldershot: Edward Elgar.

McIntyre, R. (1992). Consumption in contemporary capitalism: Beyond Marx and Veblen. *Review of Social Economy*, *Spring*, 50–57.

Moyle, B. D., McLennan, C. L. J., Ruhanen, L., & Weiler, B. (2014). Tracking the concept of sustainability in Australian tourism policy and planning documents. *Journal of Sustainable Tourism*, *22*(7), 1037–1051.

Nordic Council of Ministers. (2019). *Nordic tourism policy analysis.* Nord 2019:008. ISBN 978-92-893-6059-3 (pdf).

Nordic Council of Ministers. (2019). *Plan for Nordic cooperation 2019–2023.* PolitikNord 2019:727. ISBN 978-92-893-6158-3 (pdf).

Norwegian Ministries. (2017). *Norway's Arctic strategy—Between geopolitics and social development.* Oslo, Norway: Norwegian Ministries.

Olsen, L. S., Berlina, A., Junsberg, L., Mikkola, N, Roto, J., Rasmussen, R. O., & Karlsdottìr, A. (2016). *Sustainable business development in the Nordic Arctic* (Working Paper 2016:1). Stockholm, Sweden: Nordregio.

Pearce, P. L., & Lee, U.-I. (2005). Developing the travel career approach to tourist motivation. *Journal of Travel Research*, *43*(3), 226–237.

People's Republic of China, The State Council. (2018). *China's Arctic policy.* Beijing: People's Republic of China, The State Council. Last updated January 26, 2018.

Phillips, W. J., & Back, K. J. (2011). Conspicuous consumption applied to tourism destination. *Journal of Travel & Tourism Marketing*, *28*(6), 583–597.

Prime Minister's Office. (2015). *Growth from the North. How can Norway, Sweden and Finland achieve sustainable growth in the Scandinavian Arctic?* Report of an independent expert group. Helsinki: Prime Minister's Office Publications 04/2015.

Prime Minister's Office, Finland. (2013). *Finland's strategy for the Arctic region*. Prime Minister's Office Publications 16/2013. Helsinki: Prime Minister's Office.

Rokka, J., & Canniford, R. (2016). Heterotopian selfies: How social media destabilizes brand assemblages, *European Journal of Marketing*, *50*(9/10), 1789–1813.

Svensk Turism, A. B. (2010). *National strategy for Swedish tourism 2020/Nationell strategi för svensk besöksnäring, Strategi 2020*. Skövde: Rolf Tryckeri AB.

Swedish Lapland Visitors Board. (2020). About Swedish Lapland/Om Swedish Lapland. Retrieved from https://www.swedishlaplandvisitorsboard.com/om-destinationen/om-swedish-lapland-visitors-board/om-swedish-lapland/

Trigg, A. (2001). Veblen, Bourdieu, and conspicuous consumption. *Journal of Economic Issues*, *35*(1), 99–115. doi:10.1080/00213624.2001.11506342

United Nations World Tourism Organization (UNWTO). (2019). *Guidelines for success in the Chinese outbound tourism market*. Madrid: UNWTO. doi:10.18111/9789284421138

Zhang, C. X., Pearce, P., & Chen, G. (2019). Not losing our collective face: Social identity and Chinese tourists' reflections on uncivilised behaviour. *Tourism Management*, *73*, 71–82.

4 Chinese tourism in the Russian Arctic and Far East

Implications of a China-Russia political partnership

Mariia Kobzeva

Introduction

The new dynamic of the China-Russia political partnership invigorates all spheres of cooperation, and tourism is one of the most impressive examples. After sanctions, when Russia lost its attractiveness for visitors from Europe, Chinese visitors topped the inbound tourist list among many individual European or Asian states (excluding the Post-Soviet countries, whose statistics include migration flow) See Figure 4.1. Now, the Chinese are the most frequent guests both to the Russian Arctic and the Far East—two remote areas that have become a new priority for Russian policy post 2014. Comparison of data from 2014 and 2019 between January to September shows a 2.2% increase in numbers of Chinese tourists to Russia (Federal Agency for Tourism [Rosturizm], 2019a). Yet only 1% of them visited the Arctic area; this increase represents a valuable change for Russian regions (Ministry of the Russian Federation for the Development of the Far East and the Arctic, 2020).

A political impetus was the main factor that shaped the emergence of a Chinese market. This chapter, thus, presents empirical evidence for the argument stated in Chapter 1 of this book that Arctic tourism is an epitome of Arctic politics. However, it is fair to say that bilateral cooperation, in the Far East specifically, has chronically stagnated in spite of political fanfare (Zuenko, 2018). In this regard, it is worth examining other drivers for Chinese tourists, who are becoming interested in the Russian Arctic and the Far East. One driver is the epistemic communities, upon which this chapter directs its focus. The theory states that epistemic communities often appear in areas of cooperation with a high level of uncertainty: there is an acute need for a new level of knowledge and standards. A lack of effective official cooperation mechanisms also contributes to the informal interaction of epistemic communities (Cross, 2013). At the same time, epistemic communities are usually able to survive when they strengthen political coalitions either in the country or between states (Haas, 2008). Thus, the study of epistemic communities' formation and work helps our understanding of Arctic tourism as part of political trends in the changing Arctic.

The influx of Chinese tourists to the Russian Arctic and the Far East provided the circumstances for the formation of epistemic communities. The tourism that

DOI: 10.4324/9781003039518-4

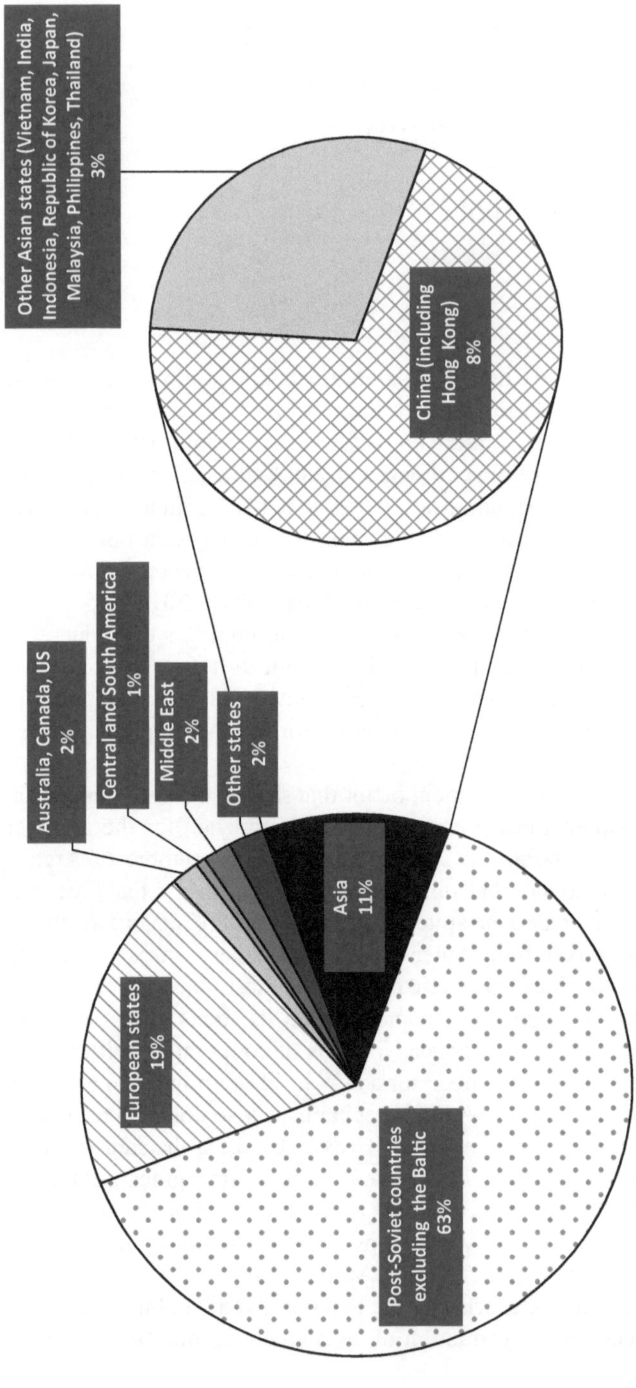

Figure 4.1 Inbound tourism to Russia in 2019 (Rosturizm, 2019a)

this chapter studies and discusses takes place in areas which are very different in terms of bilateral experience. While the Far East is tightly bounded by cross-border relations and presents a traditional destination, the Russian Arctic is a novel area for visitors and hosts (Sevastianov, 2020). New tourist trends reveal a great lack of expertise—both in organising Arctic tourism and in welcoming Chinese travellers (Sevastianov, 2020). The two states had to solve these expertise issues in a prompt way, yet they did not have a well-established knowledge to do that. This promoted the emergence of epistemic communities. They evolved from scratch into structure and united all the actors of tourism development: state actors, travel industries, and individual entrepreneurs. In this chapter, the author tries to determine—how does the Sino-Russian political partnership influence epistemic communities involved in tourism in the Russian Arctic and the Far East?

The chapter (a) evaluates Sino-Russian bilateral cooperation in tourism in the Russian Arctic and the Far East; (b) examines trends of consumption among Chinese tourists; and (c) considers shared knowledge regarding problems of cooperation and the responses of epistemic communities to them.

Data sources for the chapter

Three sources of data inform this chapter and the materials for research include a) official documents and statements published on the websites of Ministries and state agencies of Russia, b) scholarly publications both in English and Russian languages, as well as c) the author's personal consultations with representatives of the Chinese travel industry. The author held informal conversations with industry representatives. These discussions covered issues regarding the arrangement of Chinese travel to the Russian Arctic in three segments: luxury, mass tourism, and individual travel. We spoke about the preferences of Chinese tourists and the specific organisation of such tours, including logistics and funding.

The role of Chinese tourism in the Russian Arctic and the Far East regions and its impact on bilateral cooperation with Russia

For Russia, Chinese tourism is considered a domestic business, and the key point here is that it spills into a strategically important area. In geographic terms, the Russian Arctic shore spreads to the Far East that jointly forms a natural border to Europe, America, and Asia. This is literally the strongest concern for the state: to secure and develop these territories (Kobzeva, 2020; The Site of the President of Russia, 2020). Since inbound tourism contributes to regional budgets, recently the Russian government has put efforts into its support, including the development of regulatory environments, infrastructure, and business support (Ministry of the Russian Federation for the Development of the Far East and the Arctic, 2020; The Russian Government. 2019). The Federal Agency for Tourism (Rosturizm) and its working group for the development of Arctic tourism outline the goal to make the Arctic accessible for people (The Arctic, 2020a; The Arctic, 2018). New segments of tourism, including cruise and eco-tourism, are coming together regarding the adjustment of environmental standards and the re-evaluation of eco-safety of human endeavours in the hope of bringing forward solutions to the acute state tasks from a Russian government perspective (Ministry of the Russian Federation for the Development of the Far East and the Arctic, 2020; The Official Internet-Portal of Legal Information, Russian Federation, 2020).

For China, outbound tourism is considered as a part of the "soft power", propaganda and popularisation of Chinese culture abroad (Johnson, Xu, & Arlt, 2020; Lim, Ferguson, & Bishop, 2020; China Internet Information Center, 2018). As discussed in Chapter 1, one might say that Asian influences on international politics over the changing Arctic, primarily led by China, are indeed noticeably observable in the realm of Arctic tourism. This Chinese influence on the Arctic in the form of tourism is clearly captured in China's Arctic Policy White Paper of 2018 (The State Council of the People's Republic of China, 2018). In this paper, the Chinese government specifically states its interests in developing a sustainable Arctic tourism as a tourist-generating source nation to the Arctic. With this clear position and the Chinese government's Arctic-related policies, one can expect more interactions between Arctic destinations and Chinese tourists and that the latter's consumption activities would be shaped by, at least partly, the Confucian Five Constants, which were discussed in Chapter 2 of this book. This fact helps to better understand the political dimension of tourism to Russian territories. The Far East destination falls into the agenda of bilateral relations with the Russian Federation, RF, with its long history of ups and downs, intertwined with cross-border trade and migration. Even the visa-free regime for tourist groups in the area has been the case since the late Soviet Union times (Datsyshen et al., 2013). All the above shapes a solid platform for mutual understanding of prioritised security and peaceful cooperation and development of the economics of neighbouring regions.

The Arctic is a brand-new destination to the Chinese; however, it already fits into a well-formed political vision. China outlined its intention to contribute to tourism development in the High North—a process that will be connected with the Ice Silk Road initiative along the Russian Arctic and Far East shores (The State Council of the People's Republic of China, 2018). Tourism here works for the image of the state as an Arctic actor welcomed in the region. For this reason, the government promotes tourism to the Arctic explicitly and implicitly via generating interest in the High North such as popularisation of "Ice-and-Snow tourism", winter sports in the Arctic village or the "Chinese North Pole" at Mohe, in Heilongjiang province, reindeer sledding at Evenki settlement in Inner Mongolia, the chain of Polar parks established all over the country (Sa, 2009).

Against such a backdrop, the bilateral partnership supports cooperation in tourism by a range of various documents, initiatives, and working mechanisms. Starting from 1993, the agreements and memoranda focused on cross-border tourism and the key directions of visiting Moscow and St. Petersburg (Base.Garant.Ru, 1993; Rosturizm, 2019b). Currently, the list of places has expanded. The Joint Statement of 2018 mentioned the Arctic as one of the new important destinations (The Site of the President of Russia, 2018).

These documents eased travelling significantly. In particular, Chinese tourist groups from five to 50 people may visit Russia with no visa (The Ministry of Foreign Affairs of the Russian Federation, 2013; Base.Garant.Ru, 1993; Base.Garant.Ru, 2000). Cruise ships with foreigners have permission to visit an extended list of Arctic ports (Ministry of the Russian Federation for the Development of the Far East and the Arctic, 2020). The two states formed the joint list of tour operators from China and Russia, which provides services for both states (Rosturizm, 2017).

Constant foci of bilateral cooperation are raising the quality and safety of tourist services, price control, and security issues related to border traffic. The last one is especially important for both states and motivates arrangements for closer cooperation between agencies. Sino-Russian cooperation in tourism is taking place between high-level bodies and a range of bilateral mechanisms such as the Subcommittee on cooperation in the field of tourism associated with the Russian-Chinese Commission for Humanitarian Cooperation; the working group on the development and exploration of the Arctic as a part of the Sub-Commission on Trade and Economic Cooperation; Intergovernmental Commission for Cooperation of the Northeast and the Far East and Baikal Region of Russia; twin-cities collaboration mechanisms, etc. Such long-lasting collaboration provides mutual awareness.

Tourist projects have become the new sphere for attracting investment in Russian territories, ensuring employment, and providing an economic impetus for previously degraded areas. This investment also meets Chinese interests. The Northeast provinces get the chance to develop cross-border politics with Russia, which traditionally dominates their politics. In addition, they catch a new political wave by participating in Arctic national politics.

Chinese tourists' consumption in the Russian Arctic and the Far East: insights from current practices

When we talk about consumption in tourism, we mean a set of behavioural practices and preferences as well as the organisation of tourists' activities (Kirillova et al., 2020). The Arctic satisfies leisure needs related to authenticity as well as a contrast to normal routines (Urry, 2016). It offers discoveries for those who travel in groups and provides hidden and unique experiences for individuals and also to luxury tourists (Voase, 2006). Based on current practices, the preferences of Chinese visitors favour the Russian Arctic and the Far East, which are discussed in the following section. The discussion focus on the current three main types of consumption: luxury, mass, and individual consumption.

Luxury consumption: cruise and ecology recreation

Tourists interested in luxury consumption are mostly wealthy people over 50, travelling alone or with families, who come from Beijing, Shanghai, or Hong Kong. They represent a majority (33%) of participants in luxury cruise voyages travelling directly to the Russian Arctic National Park or coupled with a visit to the North Pole (Russian Arctic National Park, 2018). When travelling to the Far East, en-route tourists visit Wrangel Island, Kamchatka, and the Kuril Islands. Most Arctic voyages are provided by Poseidon Expeditions (Cypriot company, Russian owner) and Quark Expeditions (US owner) in collaboration with the Chinese travel company Jizhimei. The key vessel that brings Chinese tourists to the Russian Arctic is the Sea "50 Let Pobedy" icebreaker.

These tours are premised on two ideas. First is to achieve a unique experience of reaching the top of the world and the most remote Arctic wildlife exotic lands (Cai et al., 2020). This correlates with the educational module of cruises that includes lectures on Arctic nature and climate change. The second idea is to participate in a tour, which is not affordable for the majority of Chinese people. This includes both luxury entertainment onboard as well as a new interest in ecology consumption. Again, both of these are a sign of prosperity for the Chinese. Recreation close to clean Russian nature in the Far East forced the development of a recreation zone focused on high-quality unique tourist products such as helicopter trips for sightseeing Kamchatka volcanoes or fishing in remote crystal water rivers both in the Russian Arctic and the Far East, living with Nenets families in chums during their migration, hunting bears in Kamchatka, and buying whale meat.

Chinese investors are tracking this trend and are becoming involved in the market. One example is the construction of the "Asian village", a 650 hectares resort area in Kamchatka. The Hong Kong company, Arctic Ocean Holding Group is going to invest 81.9 B rubles into this project. This will go with direct state support from the Russian side due to Kamchatka's status as a territory of priority development. Among other things, that means that the Russian government will build infrastructure to support the tourist object (The Press Office of the Far East Investment and Export Agency, 2020).

However, it is worth mentioning that luxury tourism involves less interactions with Russian Arctic inhabitants and mostly involves a rest onboard with two hours for taking exotic and high-quality pictures ashore.

Mass consumption: nature and history

The majority of tourists visit in groups as part of the no-visa group regime. These tourists are drawn from the lower-middle-class old generation from mainland China, young people in small groups, and young couples. Most of them can afford comfortable hotels. The main purpose of the visit, in addition to the unique experience of travel to the High North, is the Aurora Borealis (Pertsev, 2019). For this reason, a major influx of these tourists arrives between late December around the Judeo-Christian calendar New Year and up to the Chinese New Year (ChinaRusLaw.Com, 2010; Cai et al., 2020).

Destinations in the Russian Arctic centre around Murmansk attractions, particularly the Arkhangelsk and Yamalo-Nenets Autonomous Region (The Arctic, 2020c; Staalensen, 2017). Several Murmansk locations like Teriberka, Lovozero, the Snow Village in Kirovsk, benefited from the Aurora Borealis so that the Northern Lights became a key Murmansk brand. Leaping at the chance of being included, surrounding areas try to position themselves as the best locations to catch the Aurora and provide services, including Aurora hunting and glass-wall hotels (Pertsev, 2019).

In the Far East, the key recipient of Chinese tourists is Primorsky Krai with primarily city and recreation tourist visits. Due to active state support of tourist development, new destinations include new narratives that animate positive historical precedents of mutually beneficial cooperation in joint tour products, for example, the Great Tea Route, "Road of Memory" route, "communist" tours, etc. (Rosturizm, 2016; Gomilevskaya & Ivanova, 2019; Adashova, 2017).

Individual consumption: adventures and business

The rest, a minority, includes entrepreneurs, officials, and scholars from mainland China, HK, and Macao, who participate in research trips, as well as single and group thrill-seekers (up to twenty people in a group), individual bloggers, and environmental volunteers (Imamova, 2017). For those adventurers, the most interesting thing in the Russian Arctic and remote Far East areas is that they were rarely visited earlier. Unknown, extreme, and wildlife areas attract this most adaptive part of Chinese tourists. Mostly, they are young people who speak English and tend to stay in hostels or Airbnb (Travel Association "World without Borders", 2014). They get the most of the cultural experience, learn about Indigenous peoples' lives, and are willing to try dog sledding in Kirovsk and reindeer sledding based in the Sami village in Lovozero, etc.

Shared knowledge and the response of epistemic communities

From the very start of raising inbound tourism from China, the Russian travel industry has faced challenges. The new travel directions were not organised well enough and were not ready to meet the demand of the rising number of Chinese clients (Sevastianov, 2020). Gradually, an understanding of the key issues of Chinese tourism in the areas formed.

The main issues were economic ones. These included issues of business ownership and profit for locals. Chinese tourists still prefer to use Chinese services to the point of non-legal hotel ownership (Potapova & Suhodolov, 2019). One of the reasons for this is the language barrier. Additionally, there are internal Russian factors that cause problems, specifically, poor local management and excessive bureaucracy. For instance, some areas in the Russian Arctic and Far East require special permission to enter and to take photos and videos. This reduces the benefit for Russian regions, both for local and Indigenous populations, and fuels fear that "the Chinese will buy up everything".

Another important issue is ecological risks. These arise from poor domestic mechanisms for protecting the fragile Arctic nature and from negative precedents with Chinese tourists in other regions, in particular around the Baikal Lake

(Karnyshev & Ivanova, 2019). There is an unsolved local problem of unregulated trash dumping in several cases which were connected with Chinese citizens' behaviours (Melnikova, 2017). Subsequently, there is an urgent need to adjust environmental standards for the tourism industry in order to ensure ecological resilience.

The utmost practical issue for the Russian tourist industry that works with Chinese visitors is the language barrier. It limits the Russian travel agencies' presence in the Chinese market. Most of the Russian websites, including the local administrations, do not have Chinese versions. Up until recently, directional signs and menus in Chinese and Chinese-speaking personnel were absent. Consequently, most information is provided by Chinese travel services such as Trip.com, Group Limited, and Qyer.com, including news on exhibitions in local museums.

An even more disappointing fact is the value of excursions. It is rare for a Russian tour guide to speak Chinese. In most and the best cases, Russian guides communicate using English while a Chinese tour leader translates the information to the group. To save money, the worst cases involve amateur Chinese guides who conduct excursions themselves, with poor knowledge of Russian history and culture (Kiselev, 2014). Unfortunately, the result is that Chinese tourists are disinformed and hardly know anything about the region to the point that they are not able to show the place on a map after they visited it (author's consultations with representatives of the Chinese travel industry).

Chinese clients' specific needs are another important issue. The first experience of welcoming visitors from the PRC sketched out the key preferences of guests. In particular, they hope to have Chinese dishes on the menu, Chinese programmes on TV, kettles or thermos bottles in the rooms, Chinese power plug adaptors, and credit card machines that allow the use of UnionPay cards (Travel Association "World without Borders", 2014). All these factors affect business and demand adaptation by cruise ships, hotels, and hostels.

Accumulated experience has shown the need for the whole industry to adapt to the new market. Instantly, that led to independent solutions by individual businesses, such as the prompt adaptation of the "50 Let Pobedy" icebreaker to Chinese visitors' needs (Travel Association "World without Borders", 2018). On the other hand, many problems required a more complex approach, and this promoted epistemic communities to run both in formal and informal ways.

The formal way includes experts, business, and state agencies that in most cases serve as the final authority for solving complex problems, for instance, the infrastructure problem. Many potential tourist destinations are simply not safe in terms of low accessibility. Hence came the business ideas to use Russian military airfields for tourist planes or to restructure the industrial zone into a premium class resort area for international cruises, such as the Nornikel project in Liinakhamari, Murmansk region (The Arctic, 2020b; TASS, 2020). This surely requires regulations and safety adjustments (Pashkevich & Stjernström, 2014).

Another example is the involvement of the Indigenous population in tourism (Jørgensen & Bertelsen, 2020). On the backdrop of lower interest in their own culture, especially among the youngest generation, such activities could serve as a motivator for keeping the traditions and unique lifestyle alive. With ecology reaching the top of agendas, Russia hopes to develop an environmentally enlightening segment of tourism and partially solve the problems of Indigenous peoples. If further developed, this will make local precedents of epistemic communities where Indigenous peoples have strong participation (Gorbuntsova et al., 2019).

Best tourism-related practice in the official epistemic community of the Russian Arctic and the Far East associated with Chinese visitors is the China Friendly project registered in Federal Technical Regulation and Metrology Agency, RF. China Friendly provides recommendations to Russian entrepreneurs, conducts assessments of their business cultures, and certifies those who are willing to improve their service (China Friendly Program, n.d.). The initiative accumulates knowledge, shares experiences between Russian and Chinese partners, and involves experts and businesspeople from the Russian Arctic and the Far East. As a result, the hotel chain "Azimut", especially in Murmansk and Vladivostok, has already achieved certification, which positions them well with Chinese travel agencies as the most advisable locations.

At the same time, a good part of Chinese epistemic communities tends to stay informal. This very dynamic and creative sphere still relies on enthusiasts whose knowledge about the Arctic remains unique in China. Against a backdrop of state support to Arctic science, more Chinese scholars, who have never been to the Arctic, hope to get a chance to explore the Arctic. One of the options for them is to participate in private tours sponsored by the wealthiest Chinese people, who seek fresh experiences. Those individual expeditions use private consultants who are able to ensure safety during travel (author's consultations with representatives of the Chinese travel industry).

Conclusion

The development of Chinese-Russian epistemic communities in tourism in the Russian Arctic and the Far East benefited when the need for new knowledge and changing standards suited political agendas. A China-Russia political partnership provided a variety of cooperation mechanisms focused on tourism. Thus, Russian tourism business benefited significantly from sharing knowledge between Chinese and Russian experts, officials, and entrepreneurs. The best example is the China Friendly project that involves entrepreneurs from the Russian Arctic and the Far East, among others.

Epistemic communities manage to solve organisational problems but still are challenged to balance consumption. In principle, tourism may be the most effective way to raise mutual awareness of each other among guests and hosts. Chinese tours allow visitors to experience the Russian Arctic and Far East and allow Russian people to get a personal experience of Chinese people. Formally, this promotes the fact that the Chinese become welcomed guests of the High North, which corresponds with the political intentions of the PRC. In practice, however, Chinese visits lack full cultural exchanges. Consumption preferences of Chinese tourists still have low cultural interest. In turn, mental and behavioural differences coupled with language barriers and insufficient educational components of tours lead to the fact that proper cultural exchange rests on enthusiasts. Finally, the economic and ecological challenges that go along with Chinese tourism with no due regard may set the Russian public against further development of the tourist industry in two areas.

The Russian tourist industry still works better for English-speaking visitors and, in general, is oriented to tourism from the Western part of the world; however, it has shown adaptability to the new Chinese/Asian focus as well. At the same time, the industry depends on relations between the states since these shape the preferences of tourists. The need for the harmonisation of Arctic tourism will heighten after the COVID-19 outbreak and require international decisiveness for cooperation. An effective means to address this is to use the capacity of international epistemic communities that face similar challenges.

References

Adashova, T. (2017). Russian Far East: Specific features of the development of development of international tourism. *Tourism and Regional Development*, *10*, 128–132. (In Russian).

Base.Garant.Ru. (1993). *Agreement between the Government of the Russian Federation and the Government of the Peoples' Republic of China on cooperation in the field of tourism (Beijing)*. Retrieved from http://base.garant.ru/2561916/ (In Russian).

Base.Garant.Ru. (2000). *Agreement between the Government of the Russian Federation and the Government of the People's Republic of China on visa-free group tourist trips (as amended)*. Retrieved from http://base.garant.ru/2562816/ (In Russian).

Cai, Y., Ma, J., & Lee, Y.-S. (2020). How do Chinese travelers experience the Arctic? Insights from a hedonic and eudaimonic perspective. *Scandinavian Journal of Hospitality and Tourism*, *20*(2), 144–165. doi:10.1080/15022250.2020.1744185

China Friendly Program. (n.d.). *China friendly*. Retrieved from http://chinafriendly.ru/#partnerscf (In Russian).

China Internet Information Center. (2018). *Minister of culture and tourism Luo Shugang: Promoting Chinese culture to the world*. Retrieved from http://russian.china.org.cn/exclusive/txt/2018-03/27/content_50753400.htm (In Russian).

ChinaRusLaw.Com. (2010). *Program of cooperation between the regions of the Far East and Eastern Siberia of the Russian Federation and the North-East of the People's Republic of China (2009–2018)*. Retrieved from http://www.chinaruslaw.com/RU/CnRuTreaty/004/201035210624_735729.htm (In Russian).

Cross, M. K. D. (2013). Rethinking epistemic communities twenty years later. *Review of International Studies*, *39*(1), 137–160.

Datsyshen, V., Larin, V., & Romanova, G. (2013). The border's lock is opened. Border regions in Russian-Chinese relations. In A. V. Lukin (Ed.), *Russia and China. Four centuries of interaction* (pp. 409–513). Diplomatic Academy, Ministry of Foreign Affairs of Russia. Retrieved from https://www.elibrary.ru/item.asp?id=25939821 (In Russian).

Gomilevskaya, G., & Ivanova, D. (2019). Cross-border as a factor in the development of international tourism in the Russian Far East. Territory of new opportunities. *Vladivostok State University of Economics and Service Bulletin*, *11*(3), 63–81. doi:10.24866/VVSU/2073-3984/2019-3/063-081 (In Russian).

Gorbuntsova, T., Dobson, S., & Palmer, N. (2019). Diverse geographies of power and spatial production: Tourism industry development in the Yamal Peninsula, Northern Siberia. *Annals of Tourism Research, 76*, 67–79. doi:10.1016/j.annals.2019.03.006

Haas, P. (2008). Epistemic communities. In D. Bodansky, J. Brunnée, & E. Hey (Eds.), *The Oxford handbook of international environmental law* (pp. 792–806). Oxford: Oxford University Press. doi:10.1093/oxfordhb/9780199552153.013.0034

Imamova, A. (2017). Bely Island was cleaned from garbage: How volunteers work in the Arctic (2017, February 7). *TASS*. Retrieved from https://tass.ru/arktika-segodnya/3992783 (In Russian).

Johnson, P. C., Xu, H., & Arlt, W. G. (2020). Outbound Chinese tourism: Looking back and looking forward. *Journal of Policy Research in Tourism, Leisure and Events, 12*(1), 1–7. doi:10.1080/19407963.2018.1505098

Jørgensen, M. T., & Bertelsen, R. G. (2020). Chinese tourism in the Nordic Arctic – Opportunities beyond the economic. *Scandinavian Journal of Hospitality and Tourism, 20*(2), 166–177. doi:10.1080/15022250.2020.1744186

Karnyshev, D., & Ivanova, E. (2019). Etymological and psychological-economic manifestations of Chinese patriotism in the Baikal region. Irkutsk State University Bulletin. Series: Political Science. *Religious Studies, 30*, 75–84. (In Russian).

Kirillova, K., Wang, D., Fu, X., & Lehto, X. (2020). Beyond "culture": A comparative study of forces structuring tourism consumption. *Annals of Tourism Research, 83*, 102941. doi:10.1016/j.annals.2020.102941

Kiselev, D. (2014). Arctic tourism in the context of China's polar policy. *Society and State in China, 44*(1), 300–306. (In Russian).

Kobzeva, M. (2020). Strategic partnership setting for Sino-Russian cooperation in Arctic shipping. *The Polar Journal, 10*(2), 334–352. doi:10.1080/2154896X.2020.1810956

Lim, D. J., Ferguson, V., & Bishop, R. (2020). Chinese outbound tourism as an instrument of economic statecraft. *Journal of Contemporary China, 29*(126), 916–933. doi:10.1080/10670564.2020.1744390

Melnikova, A. (2017). Chinese threat: Do tourists from the Middle Kingdom kill the Baikal? *RIA Novosti*. Retrieved from https://ria.ru/20171115/1508809182.html (In Russian).

Ministry of the Russian Federation for the Development of the Far East and the Arctic. (2020). *The Ministry for the Development of the Russian Far East sets a goal to increase tourist flow in the Arctic to 3 million people a year in 15 years*. Retrieved from https://minvr.gov.ru/press-center/news/25374/ (In Russian).

Pashkevich, A., & Stjernström, O. (2014). Making Russian Arctic accessible for tourists: Analysis of the institutional barriers. *Polar Geography, 37*(2), 137–156. doi:10.1080/1088937X.2014.919040

Pertsev, A. (2019, December 30). Conceive under the northern lights. Due to Eastern beliefs, the number of Chinese tourists in Murmansk has grown 100 times. *Meduza*. Retrieved from https://meduza.io/feature/2019/12/30/zachat-pod-severnym-siyaniem (In Russian).

Potapova, E., & Suhodolov, V. (2019). Harmonization of recreational opportunities and a growing tourist flow in the Baikal natural territory. *Bulletin of the Baikal State University, 29*(1), 7–17. (In Russian).

Russian Arctic National Park. (2018). *Statistics*. Retrieved from http://www.rus-arc.ru/ru/Tourism/Statistics (In Russian).

Sa, G. (2009). Welcome to Genhe. *Scribd*. Retrieved from https://www.scribd.com/document/137921345/Welcome-to-Genhe

Sevastianov, D. (2020). Arctic tourism in the Barents Sea region: Current state and potential limits. *Arctic and North*, *39*, 26–36. (In Russian).

Staalensen, A. (2017, March 16). More Chinese tourists choose Murmansk. *The Independent Barents Observer*. Retrieved from https://thebarentsobserver.com/en/travel/2017/03/more-chinese-tourists-choose-murmansk

TASS. (2020, February 28). Nornickel will invest 2.3 billion rubles in the development of the Murmansk region in the next three years. *TASS*. Retrieved from https://tass.ru/ekonomika/7868177

The Arctic. (2018). *Rosturizm created a working group for the development of tourism in the Arctic*. Retrieved from https://ru.arctic.ru/tourism/20181101/799219.html (In Russian).

The Arctic. (2020a). *Head of Rosturizm: We need to make the Arctic more accessible to people, but at the same time preserve its unique nature*. Retrieved from https://ru.arctic.ru/tourism/20200918/978750.html (In Russian).

The Arctic. (2020b). *Tourism future of the Arctic discussed in Murmansk*. Retrieved from https://ru.arctic.ru/tourism/20200923/980109.html (In Russian).

The Arctic. (2020c). *Yamalo-Nenets Autonomous Okrug, Murmansk and Arkhangelsk regions became leaders in tourist flow among the Arctic regions of Russia*. Retrieved from https://ru.arctic.ru/tourism/20200604/946049.html (In Russian).

The Federal Agency for Tourism, (Rosturizm). (2016). *Russia, Mongolia and China will develop a joint travel brand 'Great Tea Route'*. Retrieved from https://www.russiatourism.ru/news/10837/ (In Russian).

The Federal Agency for Tourism, (Rosturizm). (2017). *Tour operators carrying out visa-free travel in the Chinese direction*. Retrieved from https://www.russiatourism.ru/contents/deyatelnost/mezhdunarodnaya-deyatelnost/mezhregionalnoe-i-dvustoronnee-sotrudnichestvo/turoperatory-osushchestvlyayushchie-bezvizovye-poezdki-na-kitayskom-napravlenii/ (In Russian).

The Federal Agency for Tourism, (Rosturizm). (2019a). *Rosturizm signed a memorandum on cooperation in the field of tourism between Russia and China*. Retrieved from https://www.russiatourism.ru/news/16422/ (In Russian).

The Federal Agency for Tourism, (Rosturizm). (2019b). *Selected statistical information calculated in accordance with the Official Statistical Methodology for estimating the number of inbound and outbound tourist trips*. Retrieved from https://www.russiatourism.ru/contents/statistika/statisticheskie-pokazateli-vzaimnykh-poezdok-grazhdan-rossiyskoy-federatsii-i-grazhdan-inostrannykh-gosudarstv/vyborochnaya-statisticheskaya-informatsiya-rasschitannaya-v-sootvetstvii-s-ofitsialnoy-statisticheskoy-metodologiey-otsenki-chisla-vezdnykh-i-vyezdnykh-turistskikh-poezdok/ (In Russian).

The Ministry of Foreign Affairs of the Russian Federation. (2013). *Agreement between the Government of the Russian Federation and the Government of the People's Republic of China on facilitating citizens' travel*. Retrieved from https://www.mid.ru/foreign_policy/international_contracts/2_contract/-/storage-viewer/bilateral/page-1/44427?_storageviewer_WAR_storageviewerportlet_advancedSearch=false&_storageviewer_WAR_storageviewerportlet_keywords=%D0%BA%D0%B8%D1%82%D0%B0%D0%B9+2001&_storageviewer_WAR_storageviewerportlet_fromPage=search&_storageviewer_WAR_storageviewerportlet_andOperator=1 (In Russian).

The Official Internet-Portal of Legal Information, Russian Federation. (2020). *Decree of the President of the Russian Federation dated 05.03.2020 No. 164, "On the basics of state policy of the Russian Federation in the Arctic for the period until 2035"*.

Retrieved from http://publication.pravo.gov.ru/Document/View/0001202003050019?index=1&rangeSize=1 (In Russian).

The Press Office of the Far East Investment and Export Agency. (2020, September 1). *The first stage of creation of the 'Asian Village' tourist cluster begins in Kamchatka*. Ministry of the Russian Federation for the Development of the Far East and the Arctic. Retrieved from https://minvr.gov.ru/press-center/news/27767/ (In Russian).

The Russian Government. (2019). *The strategy for the development of tourism in Russia until 2035 was approved*. Retrieved from http://government.ru/docs/37906/ (In Russian).

The Site of the President of Russia. (2018). *Joint statement of the Russian Federation and the People's Republic of China*. Retrieved from http://kremlin.ru/supplement/5312 (In Russian).

The Site of the President of Russia. (2020). *Decree of the President of the Russian Federation "On the strategy for the development of the Arctic Zone of the Russian Federation and ensuring national security for the period until 2035"* (26.10.2020). Retrieved from http://static.kremlin.ru/media/events/files/ru/J8FhckYOPAQQfxN6Xlt6ti6XzpTVAvQy.pdf (In Russian).

The State Council of the People's Republic of China. (2018). *Full text: China's Arctic policy*. The White Paper. Retrieved from http://english.www.gov.cn/archive/white_paper/2018/01/26/content_281476026660336.htm

Travel Association 'World without Borders'. (2014). *Infographic # 3: 'What are the peculiarities of receiving Chinese tourists in hotels?'*. Travel Association 'World without Borders'. Retrieved from https://www.visit-russia.ru/news/infografika-3-kakovy-osobennosti-priema-kitayskih-turistov-v-gostinicah (In Russian).

Travel Association 'World without Borders'. (2018). *Chinese tourists are more and more attracted by tours to the Arctic and Antarctic*. Travel Association 'World without Borders'. Retrieved from https://www.visit-russia.ru/news/turistov-iz-kitaya-vse-bolshe-privlekayut-tury-v-arktiku-i-antarktiku (In Russian).

Urry, J. (2016). The 'consumption' of tourism. *Sociology*, *24*(1), 23–35. doi:10.1177/0038038590024001004

Voase, R. (2006). Creating the tourist destination: Narrating the 'undiscovered' and the paradox of consumption. In K. Meethan, A. Anderson, & S. Miles (Eds.), *Tourism, consumption and representation: Narratives of place and self* (pp. 284–300). Wallingford: CABI. Retrieved from http://ebookcentral.proquest.com/lib/tromsoub-ebooks/detail.action?docID=289458

Zuenko, I. (2018). *A milestone, not a turning point: How China will develop the Russian far east*. Carnegie Moscow Center. Retrieved from https://carnegie.ru/commentary/77671

5 Greenland, the Faroe Islands, and Chinese tourism

Sustainable development in the face of otherness, growth, and power

Nathan Reigner, Gestur Hovgaard, and Firouz Gaini

Introduction

Greenland and the Faroe Islands are on the frontier of mainstream tourism in the Arctic. They are islands in the North Atlantic Ocean with harsh climates, dramatic scenery, small populations, and relatively limited accessibility (i.e. a limited range of air and sea connections) see Figure 5.1. As tourism in the Arctic has grown and nearby destinations like Iceland have become more saturated, tourists seeking lesser-known and lesser-developed (relative to tourism) destinations in the west Nordic region are visiting Greenland and the Faroe Islands. Tourists from China are no exception.

Both Greenland and the Faroe Islands have appreciated the economic benefits of increasing tourism and, each in their own way, are working to develop their tourism sectors sustainably. To leverage tourism as a sustainable development tool, these countries need to build their tourism industries, infrastructure, and administration in ways that serve the needs and desires of community residents and visiting tourists simultaneously. These goals can sometimes be in tension with each other and neither communities nor tourists are static or monolithic. Therefore, in aiming to sustainably develop the inbound Chinese tourist market, particular contexts (i.e. geography, culture, politics) of Chinese tourism vis-à-vis Greenland and the Faroe Islands must be specifically taken into account.

In this chapter, the main question we seek to address is, what should tourism managers in Greenland and the Faroe Islands consider to sustainably develop their tourism sectors in the coming decade, particularly with reference to Chinese tourist visitation. We address this question by introducing Greenland, the Faroe Islands, and their respective tourism; reviewing what is known about Chinese tourism to these destinations; discussing aspects of tourism development and sustainability as they relate to the particular situation of Chinese tourism development in Greenland and the Faroe Islands; and concluding with constructive considerations for tourism development in these places.

These considerations, and the synthesis that leads to them, are based on conceptual and theoretical foundations that position outcomes that are desirable to

DOI: 10.4324/9781003039518-5

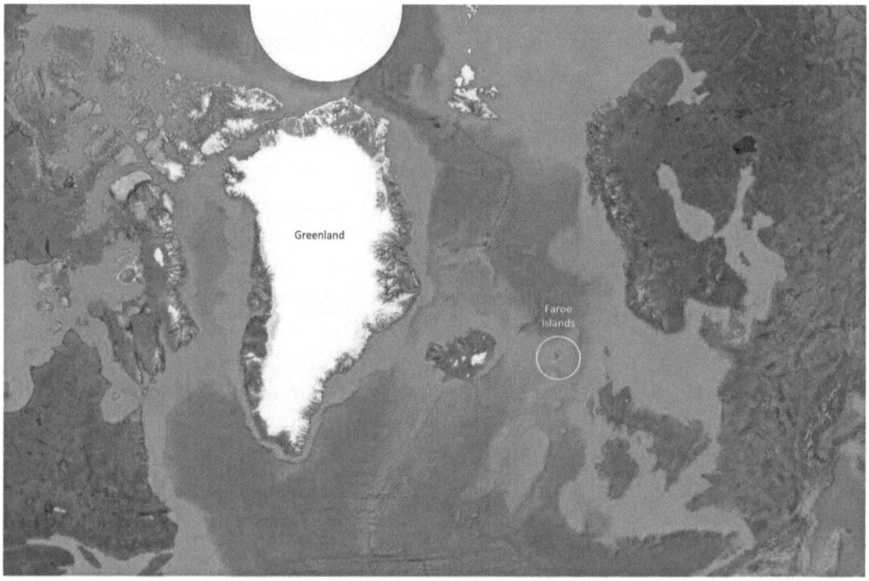

Figure 5.1 Map of Greenland and Faroe Islands

and in service of destination communities as the primary goal of development. Such a framing is prevalent in both the literature of academics and in the real work of tourism development and management, with examples proliferating at multiple scales (e.g., Cahill et al., 2018; EFLA, 2019; Gössling et al., 2012; Ren et al., 2021). This framing is particularly important for tourism development that seeks to achieve local social, environmental, and economic sustainability goals in a tourism system marked by global influences and power imbalances.

We are writing this chapter during the COVID-19 pandemic, which has dramatically interrupted normal and predictable travel and tourism patterns. This disruption has affected both inbound tourism to Greenland and the Faroe Islands and outbound tourism from China. At the time of writing, we cannot predict when the pandemic will end, if travel and tourism patterns and trends will resume where they left-off at the start of 2020, and, if not, what travel and tourism will be like after the pandemic is over. While this uncertainty leaves us in a difficult position vis-à-vis discussing the likely future of and making specific recommendations for sustainable development of tourism, we do think that there are meaningful influences and consistent principles that will be equally applicable in any post-COVID-19 future.

With this understanding, we have largely omitted references to the COVID-19 pandemic (and its effects on travel and tourism) from this chapter. When we discuss statistics, patterns, and trends we do so using the assumption that they will continue on recent trajectories into the future as they were in the years preceding the pandemic. We do this for simplicity, to limit the time and effort we spend speculating about the unknowable post-COVID-19 future, and to focus our discussion on the influences and principles we think have significant independently from the COVID-19 pandemic.

Additionally, it is worth highlighting that the effect the COVID-19 pandemic has, is having, and will have on the global tourism industry is, itself, an illustration of the importance of sustainable development principles for tourism. The sudden and unexpected cessation of virtually all international tourism revealed the kinds of dependencies and vulnerabilities that can lead to unsustainable and potentially failed development. In this chapter, we discuss sustainable tourism development as it relates to the growing inbound Chinese tourism to Greenland and the Faroe Islands. These thoughts and suggestions, as well as those that will surely follow from the industry's overall COVID-19 experience, should reinforce each other to highlight the ever-present need to develop tourism with consistent and thorough adherence to sustainability principles.

Greenland and the Faroe Islands

Greenland and the Faroe Islands have both many similarities and significant differences. They are both nations in the North Atlantic Ocean. Each has a harsh, charismatic climate and dramatic, compelling land- and seascapes. Both are self-governing member nations of the Danish kingdom. Each has approximately 55,000 residents, about one-third of whom live in each country's capital city (Nuuk in Greenland and Tórshavn in the Faroe Islands). Although their historical paths to the present have been nuanced and had multiple phases, both societies gained self-rule in the middle or late 20th century and made their transitions from subsistence economies to modern and technologically advanced ones with relative rapidity. Both countries derive substantial national revenue from marine fisheries. Furthermore, the two countries have their respective native languages (Kallalisut in Greenland and Føroyskt mál in the Faroe Islands) and Danish as a second language. English proficiency is common among both populations.

Greenland and the Faroe Islands are significantly different as well. Greenland is the largest island on Earth, supports the largest extent of glacial ice outside of Antarctica, is among the most northerly land masses, and has the most northerly communities on Earth. The Faroe Islands form an archipelago of quite small islands (1,399 km^2 in total) and are relatively temperate compared with true Arctic destinations like Greenland. Greenland's population is nearly 90% Inuit. The Faroese trace their cultural roots to Viking Europe.

In addition to these environmental and social differences, there are key contemporary political differences between Greenland and the Faroe Islands. From the 18th to early 20th centuries, Greenland was kept in the political status of a colony by Denmark, while the Faroe Islands were considered a county within the Danish state. While both countries were subject to trade monopolies imposed by Denmark, the Faroe Islands were released from theirs in 1856 while

Greenland and its Inuit population was kept generally isolated from the outside world until World War II. While both countries have similar status as self-governing member countries of the Danish kingdom, independence movements are, at present, considerably more active and prominent in Greenland than in the Faroe Islands.

During World War II and continuing into the Cold War, the geostrategic importance of Greenland to North Atlantic security became a driving force in Greenland's development and international relations. Greenland lies at the western edge of the Greenland-Iceland-United Kingdom gap, the passage between the Arctic and North Atlantic Oceans. This position, and its significance for North American and Western Europe vis-à-vis shipping, military movement, and natural resource availability lead to the establishment and occupation of several American and North Atlantic Treaty Organization (i.e. NATO) military bases and significant global security and diplomatic attention. With the Arctic warming and its ice melting, this sometimes-unwanted global attention is intensifying (e.g. Bykova, 2021; Hook, 2019; Schionning, 2020). Although similar attention is also directed at the Faroe Islands, its intensity level and the significance of its potential impact is somewhat less than for Greenland.

Tourism in Greenland and the Faroe Islands

Tourism to and in Greenland and the Faroe Island exhibits many similarities as well. These similarities span visitation, tourist experiences, destination development, and governance. Like many Arctic tourism destinations, visitation to Greenland and the Faroe Islands has been growing in recent years. In the Faroe Islands between 2011–2017, the number of hotel bed nights grew by approximately 33%, and tourism revenues grew by nearly 58% (Visit Faroe Islands, 2019b). During a similar time period in Greenland, tourism growth patterns tracked those of the Faroe Islands, with non-Greenlanders overnight hotel stays and departing flight passengers increasing by 19% and 13%, respectively (Visit Greenland, n.d.). For both destinations, ship-based tourism (i.e. cruise ships and the Smyril Line's Norrøna ferry to the Faroe Islands) represents a large and growing proportion of all tourist visitation in each country. In 2018, ship-based tourism arrivals accounted for approximately 50,000 visitors to each country, which in Greenland is approximately 240% of the number of cruise passengers arriving in 2014 (Visit Faroe Islands, 2019a; Visit Greenland, n.d.).

More and more tourists are visiting Greenland and the Faroe Islands because of the unique experiences available at the destinations, most of which are connected in some way to the islands' nature. Each country boasts dramatic scenery (e.g. fjords, glaciers, mountains, coastlines) and opportunities for an intimate connection to wildlife (e.g. whales, polar bears, puffins and other seabirds, sheep). Each brands its tourist settings and experiences as pristine, unspoiled, and authentic, offering tourists opportunities to escape familiar environs and be among the few in the world who have visited these remote and little-known destinations (Gudlaugsson & Magnússon, 2012).

While seeing and experiencing the islands' nature is the primary characteristic promoted by tourist imagery and motivating tourist interest, the connections between this nature and the islands' cultures and peoples must be acknowledged. While visiting these destinations, most tourists necessarily spend as much time in cities and villages as they do in, at least perceived, wild and unspoiled nature. During this time, tourists have opportunities to observe and interact with locals, who are themselves as individuals and communities intimately connected to the nature that draws tourists. In Greenland, the ice, sea, and wildlife that are the subject of tourists' attention are the resources upon which Greenland's Inuit population has lived for centuries and millennia (Kaae, 2002; Smed, 2014). In the Faroe Islands, the green mountains, roaming sheep, and masses of birdlife enabled the unique Faroese society to establish itself and flourish. Tourists' focus on nature in both destinations does, however, create difficulties for host communities and members of Greenlandic and Faroese society. So much attention on experiential concepts of the unspoiled, pristine, and authentic in some ways denies the existence of the destinations' modern and, for their residents', conventional societies. It establishes a dynamic in which residents, especially those who interact with tourists, are pressured to perform an often minimising and anachronistic version of themselves and their communities—one that ignores their contemporary living situation, distances themselves from their desired future, and reinforces unhelpful stereotypes of a timeless and primitive Arctic (Shackel, 2011). In the worst cases, this dynamic can lead to a "frontstage—backstage" division in which tourists are denied opportunities for the authentic experiences they seek, or at least purport to seek, and local residents are compelled, if they wish to capitalise on tourism, to present a caricature of themselves (Goffman, 1959; Shackel, 2011).

While tourism development can be fraught for host communities, growing and developing tourism is a goal for both Greenland and the Faroe Islands. Within each country, at the national level, tourism is viewed as a complementary industry to fishing, and mining in the case of Greenland, and aquaculture for the Faroe Islands (Visit Faroe Islands, 2019b; Visit Greenland, n.d.). In particular, tourism is valued as a means to develop sustainable and productive economies in the upcoming decades (Visit Faroe Islands, 2019b; Visit Greenland, n.d.). Key to realising this goal is ensuring that both the benefits and burdens of tourism are equitably distributed in both space (i.e. geographically, from capital to periphery) and time (i.e. seasonally). With a single primary international arrival point, a single dominant capital city, and an otherwise widely distributed population, both Greenland and the Faroe Islands face challenges to broadening the geographic distribution of tourism. Additionally, their northerly locations and their climatic implications (e.g. winter cold, storms, and darkness) pose inherent difficulties for accomplishing the countries' goals of enhancing four-season tourism industries (Rantala et al., 2019).

These geographic and temporal constraints on growing and developing tourism in Greenland and the Faroe Islands are accompanied by administrative and infrastructural constraints. At the strategic level, a need for more and better coordination within state government, between local and state government, and between government and industry is nearly universally cited by destination managers and tourism analysts as a critical need for the sustainability of tourism development (Petridou et al., 2019; Reigner et al., 2020; Visit Faroe Islands, 2019b; Visit Greenland, n.d.). Of particular significance within these calls for coordination is the acknowledged need in both Greenland and the Faroe Islands for administrative and legislative frameworks, potentially like those used to guide tourism development in Iceland, to be established (Ministry of Industry and Innovation, 2015). Indeed, this need for coordination and collaboration is recognised as extending beyond national boarders to the North Atlantic region (i.e. the Faroe Islands, Greenland, and Iceland) and the Nordic realm (North Atlantic Tourism Association, n.d.; Nordic Council of Ministers, 2019). At the operational level, both Greenland and the Faroe Islands need to address capacity constraints related to tourist infrastructure (e.g. lodging, airports, ports, trails) and adequately trained tourism professionals (Qu et al., 2011; Reigner, 2019; Reigner et al., 2020; Rosen, 2016).

Chinese tourism to and in Greenland and the Faroe Islands

Generally, while tourism in Greenland and the Faroe Islands is marked by consistent public, professional, and administrative attention, comparatively little attention has been devoted to Chinese tourism specifically. This situation is not unique to Greenland and the Faroe Islands; most Arctic and northern periphery tourist destinations suffer from a paucity of empirical and documented knowledge about Chinese tourists and their travel patterns and experiences (Haung et al., 2017). While this limits the information available to tourism policy makers, managers, operators, and host communities, the gap is rapidly being filled by both scientific researchers and destination management organisations. In 2019, just over 1,000 Chinese tourists visited Greenland arriving by air and accounting for approximately 1,600 overnight stays by approximately 650 guests in Greenlandic accommodations (Visit Greenland, 2020). This number of Chinese tourists arriving by air has remained relatively steady for the previous two to three years and represents approximately 2% of annual all air travel tourists (Visit Greenland, 2018). In addition, approximately 650 Chinese tourists visited Greenland in 2019 via cruise ships (Visit Greenland, 2020). Less detail is presented in official statistics of Chinese tourism in the Faroe Islands, where just over 4,000 guest nights were spent by tourists from all of Asia, representing approximately 2.5% of all guest nights (Visit Faroe Islands, 2019a).

Although statistical information about Chinese tourists is limited, growing and developing the inbound Chinese tourism market is a priority for Greenland, the Faroe Islands, and all of the Nordic realm. The co-operative plan for Nordic tourism development between 2019 and 2023 celebrates success in courting long-haul inbound tourism markets, which necessarily includes the Chinese market, and emphasises continuing and intensifying this effort in the future (Nordic Council of Ministers, 2019). This inter-governmental call for the development of the Chinese tourism market is driven, in part, by the view of Chinese tourists as an economic resource with a large and growing potential to infuse revenues into tourist industries and host communities (Bislev & Smed, 2018). Management and economic awareness of the importance of the Chinese inbound market is reflected by calls to increase research and data gathering, which is reflected in the publication of this volume and related scholarly works (Lee, 2020).

The growing body of research on Chinese tourists to Arctic and northern peripheral destinations, including Greenland and the Faroe Islands, suggests the presence of tensions or incongruities both within tourists themselves and between Chinese tourists and host communities. One point of agreement, within both the characterisation of Chinese tourists to these destinations and the destinations' strategic plans for development, is that a trip to Greenland or the Faroe Islands for a Chinese tourist is a trip to a remote, wild, adventurous, unique, powerful, and pristine destination (Bislev & Smed, 2018; Martin & Ren, 2020; Visit Faroe Islands, 2019b; Visit Greenland, n.d.). Within this framing, Chinese travel to one of these destinations, be it a cruise, a visit to urban Nuuk or Tórshavn, or a wilderness trek, stands in for a visit to the Arctic as a whole (Bislev & Smed, 2018). That is, in the minds of tourists from as distant and different a locale as China, any one particular destination is often experienced as representative of the Arctic as a whole. This transitive phenomenon can create tensions between Chinese tourists and host communities in which each population (tourist and resident) stereotypes or caricatures the other, ignoring local context and effectively distancing the two populations further (Bislev & Smed, 2018). The cognitive space opened by this distance is evidenced in the behaviours of some Chinese tourists. Although Chinese tourists to Greenland by and large embark upon their trips for adventure and a close connection with Arctic nature, the activities in which they actually participate are often less than fully adventurous (i.e. guided walking or sightseeing by boat, not wilderness treks), and Chinese tourists often arrive less than fully prepared for the Arctic conditions they purport to seek (Martin & Ren, 2020; also see chapters by Cheung, and Kotašková and Jelínek in this volume). These behaviours can be accompanied by expectations and often needs among Chinese tourists for services and facilities common in more mainstream travel destinations but uncommon in the Arctic and peripheral settings, including fast and free internet connectivity; low prices; diverse and high-quality food and beverage, including Chinese menu options; Chinese language media and tourism services (Martin & Ren, 2020). These patterns seem to hold true even among young, cosmopolitan, and independent Chinese travellers (Lojo, 2020). Just as there is a distance between the expected or projected experiences and the actual experiences of tourists, there is also a distance between the actual characteristics and desires of Chinese tourists and the assumptions or preconceived notions held by tourism operators and industry stakeholders with respect to Chinese tourists (Dichter et al., 2018). Non-Chinese operators and stakeholders often extrapolate on cultural representations of Chinese tourists and apply common stereotypes of Chinese tourists as dependent upon Chinese-specific services and unwilling to engage directly with local communities and cultures (Martin & Ren, 2020). Although there can be commonalities among Chinese tourists and the interactions between Chinese tourists and destination communities, it is important to acknowledge and incorporate the diversity that does indeed exist within the population of Chinese residents and nationals travelling to the Arctic and northern periphery (Dichter et al., 2018; McKinsey & Company, 2018).

China, tourism, and the 21st-century Arctic

Tourists are not the only members of Chinese society with a growing interest in the Arctic. Indeed, the Arctic is a central element of China's geostrategic focus, and activity and tourism can aptly be viewed as an extension of the geopolitical spheres between the Arctic and China (see Chapter 1 of this book). This strategy is organised around intertwined economic and national security goals. These goals include securing access to abundant and yet unexploited natural resources in the Arctic, establishing logistical infrastructure to leverage the "Polar Silk Road" shipping lanes opening in the Arctic Ocean as sea ice melts, and maintaining the ability to project global power in the Arctic region through both civil and military means. Greenland, in particular, is a key target of China's Arctic focus and activities (Volpe, 2020). Greenland is rich in mineral resources, which are of interest to China's manufacturing economy. It is also a potential location for transshipping between Asian origin ports and North American destination ports, and its position both at the western end of the passages from the Arctic to North Atlantic Oceans and on the transpolar flight path between Eurasia and North America is of critical national security interest. Greenland, which has both investment needs and a desire to diversify its relationships with foreign countries, could potentially benefit from China's proven willingness and ability to invest money, technical expertise, and political capital.

Tourism has emerged as one of the mechanisms China uses to leverage economic influence and power to accomplish its geostrategic goals around the world and in the Arctic (Paik, 2020). China applies tourism to accomplish its geostrategic goals with both the development and restriction of its outbound tourists (Lim et al., 2020). Using regulatory tools like the Approved Destination Status and its restrictions on group tour agencies and economic tools like investment in tourism destinations and destination marketing, China helps to develop tourism destinations. In some cases, these are geographically concentrated in locations of strategic interest to China or developed in ways that cater specifically to Chinese tourists. China has also exerted control in restrictive ways, using its regulatory authority to stop as well as nationalist appeals to discourage Chinese tourists from travelling to destinations (Erokhin, 2019). In several cases, China has combined these mechanisms (i.e. developed a tourism destination, then threatened to cut or actually cut off the supply of Chinese tourists) with significant economic and political effect (Lim et al., 2020). These dynamics have been at play in the Arctic, influencing tourism development in and around the Arctic National Park in Russia and surrounding proposed, but ultimately unsuccessful, land purchases and tourism development in Iceland (Huijbens & Alessio, 2015; Zelenskaya, 2018). China's demonstrated interest in the Arctic combines with its status as the world's biggest potential outbound tourist market to suggest that such interplay among politics, economics, tourism, and national security will be prominent in Arctic tourism for the foreseeable future.

Tourism, sustainability, and development

Although often associated with countries or regions suffering from high levels of poverty and low technical and economic capacity, the concept of sustainable development and its goals apply in fact to everyone everywhere. Sustainable development is commonly conceptualised as progress toward meeting the needs of current generations without compromising the ability of future generations to meet their own needs. As such, sustainable development is an inclusive idea; it seeks to harmonise economic, environmental, and social objectives, is focused on the well-being of both individuals and communities, and reflects the United Nations interpretation of and recommendations for tourism sustainability (United Nations General Assembly, 2015). When this definition is applied to tourism, sustainable tourism development can be thought of as tourism development that provides current community residents with the economic, social, and environmental conditions they desire, delivers future community residents a resilient, flexible, and responsive tourism sector, and provides tourists with high-quality, satisfying tourism experiences (United Nations General Assembly, 2015). A contextualised definition of sustainability like this is particularly important for relatively small societies like Greenland and the Faroe Islands, whose cultural practices and community structures are vulnerable in the face of hegemonic modernity and climate change.

Sustainable development concepts are prominently presented in the tourism strategies of both Visit Greenland and Visit Faroe Islands (Visit Faroe Islands, 2019b, Visit Greenland, n.d.). These plans target multiple sustainable development goals related to education, work, economic growth, consumption and production, communities, and partnerships. In doing so, they identify and monitor indicators of tourism sustainability and collect and report data on the economic productivity of tourism, the quality of tourist experiences, and the satisfaction of local residents with tourism in and around their communities. The satisfaction of local residents with tourism in and around their communities is a particularly salient element for the sustainability of tourism and its development. Indeed, mere satisfaction is inadequate to realise tourism development sustainability. Rather, sustainable tourism development should be driven to create and sustain the environmental, social, economic, and political benefits that residents of destination communities want from tourism, i.e. their desired conditions. This root of sustainable tourism development was succinctly expressed by an informant of the ARCTISEN project:

> I just hope that it is always the Greenlanders that define the roles, what is right and what is wrong and what to do. Right now, we also have large operators

> from outside. If Greenland wants tourism, it should be under their rules … So, we should do it from inside.
>
> (Ren et al., 2020)

Phrased more personally by Julia Pars, head of Visit Greenland: "Whom do we want to invite to our home? How should we host them?" (Ren et al., 2020).

The first steps to achieving this goal must include situating tourism development power in the hands of Greenlanders and Faroese and addressing the power imbalances that these small, at least in population and economic influence, countries face with respect to global corporate and government tourism actors (Kaiser et al., 2019). Power imbalances between host destinations and tourism companies are most prominently related to cruise tourism, particularly large-ship cruise companies. These cruises operate transnationally and in the absence of a comprehensive regulatory framework that extends beyond maritime activities to address the substantial land- and community-based interactions and impacts of their tourism activities (James et al., 2020). Visit Faroe Islands' strategic plan for tourism development recognises this power imbalance directly, calling for regulatory control over the allowable size of arriving cruise ships (Visit Faroe Islands, 2019b). Similar asymmetric power relationships exist and are, in some ways, intentionally fostered between China as a source of outbound tourists and Greenland and the Faroe Islands and destination developers and managers (Lim et al., 2020). Acknowledging these power relationships and being purposeful about addressing, mitigating, or equalising them is a necessary element for the sustainability of tourism development in Greenland, the Faroe Islands, and all Arctic tourism destinations.

Multiple mechanisms are available for empowering Greenland and the Faroe Islands and equalising the power imbalance Greenlandic and Faroes communities face with respect to large and global actors and forces. Guidelines, like those developed through the Association of Arctic Expedition Cruise Operators, establish acceptable practices and hold industry operators and stakeholders accountable to the expectations of host communities (Association of Arctic Expedition Cruise Operators, n.d.; Ren et al., 2020). Domestic regulatory mechanisms like concessions and permitting systems have also proven effective for preserving local access to resources and experiences while allowing for economic development of tourism industries like fishing, hunting, trekking, and other wild-land tourism activities (Ren et al., 2020). To be effective and support harmony both within destination resident communities and among residents, tourists, and tourism operators, these regulatory mechanisms must be coordinated in their development, transparent in the communication, and consistent in the implementation (Visit Faroe Islands, 2019b). As tourism to Greenland and the Faroe Islands increases, as we must anticipate it will, it will be important that these regulatory frameworks create, maintain, and even privilege opportunities for local investment, economic enterprise, and social involvement (Hunt & Stronza, 2014). With these opportunities, resident communities in growing tourism destinations have experienced tourism as a creator of benefits, satisfaction, and happiness. Without them, residents have experienced tourism as a negatively impacting and socially degrading force.

Preventing and mitigating conflict at the institutional and individual levels between tourism or tourists and communities or residents accompanies destination empowerment as a critical dimension of action toward sustainable tourism development. The guidelines, regulations, and economic opportunities cited above are part of this. Integrated and proactive planning is another important action that must be implemented. Tourism development in Greenland and the Faroe Islands has the potential to directly alter land- and seascapes and, consequently, interrupt land and sea uses of great cultural significance to residents (Ren et al., 2021). When considering tourism development, for the Chinese inbound market or otherwise, the interconnections between tourism and virtually all infrastructure, administrative, environmental, and social aspects of destination communities must be taken into account (Ioannides, 2019; Plieninger et al., 2018). This kind of inclusive and comprehensive planning and decision making seeks tourism development in which residents and tourists (and their infrastructure, services, administration, etc.) co-exist and mutually reinforce each other, with tourism development helping residents live the lives they desire and tourists have the experiences they seek (Martin & Ren, 2020; Ren et al., 2021). Examples of such complements range from ensuring that operational tourism development (i.e. facilities and services) are attractive to, accessible for, and used by residents to strategic prioritisation of resident interests and sensibilities over those of non-residents (e.g. foreign tourist, corporate, or national interests).

Concluding thoughts

China is working hard to solidify its position as an Arctic power, securing access to the region's natural resources, establishing trans-Arctic shipping routes, and growing its global influence. Throughout the world and in the Arctic, it uses tourism as an instrument to accomplish these geostrategic goals. Greenland and the Faroe Islands are two of the Arctic and northern periphery's most rapidly emerging tourism destinations. As such, they are the focus of much attention by tourist and tourism interests, including Chinese tourists and the Chinese national tourism apparatus.

Being small states with keen interests in developing their tourism economies, Greenland and the Faroe Islands should both welcome this interest and be cautious when developing and managing tourism. This is particularly true as it relates to the Chinese inbound market with its distinctive tourist characteristics, tourism sector, and national tourism policies. This is not to say, however, that Greenland and the Faroe Islands should not focus on developing tourism for Chinese visitors—indeed they should. Doing so will help Greenland and the Faroe Islands accomplish many of their sustainability, economic, development, and international relations goals.

Developing a strong and sustainable tourism sector, including the incorporation of Chinese inbound tourists, will diversify their relationships with foreign countries and support respective national institutions and administrative capacities to work productively with China for mutual benefit while avoiding the jeopardy some other destinations have faced (Jørgensen & Bertelsen, 2020). In doing so, Greenland and the Faroe Islands should understand sustainable tourism development, including development for Chinese tourists, as development that helps their countries realise their own desired economic, social, environmental, cultural, and political futures.

References

Association of Arctic Expedition Cruise Operators. (n.d.). Guidelines. Retrieved from https://www.aeco.no/guidelines/.

Bislev, A., & Smed, K. (2018). When will the iceberg melt? Narrating the Arctic among Chinese and Danish tourists aboard a cruise ship in Greenland. *China & the Arctic, 2018*, 29–42.

Bykova, A. (2021, January 18). Two United States-Greenland collaborations get US State Department funding. *High North News*. Retrieved from https://www.highnorthnews.com/en/two-united-states-greenland-collaborations-get-us-state-department-funding

Cahill, K., Collins, R., McPartland, S., Pitt, A., & Verbos, R. (2018, January). Overview of the interagency visitor use management framework and the uses of social science in its implementation in the National Park Service. *The George Wright Forum, 35*(1), 32–41. George Wright Society.

Dichter, A., Chen, G., Saxon, S., Yu, J., & Suo, P. (2018). *Chinese tourists: Dispelling the myths. An in-depth look at China's outbound tourist market*. Retrieved from https://www.mckinsey.com/~/media/mckinsey/industries/travel%20transport%20and%20logistics/our%20insights/huanying%20to%20the%20new%20chinese%20traveler/chinese-tourists-dispelling-the-myths.pdf.

EFLA. (2019). *Jafnvægisás ferðamála (Balancing the axes of tourism)*. Reykjavik: Ministry of Industry and Innovation. Retrieved from https://www.stjornarradid.is/default.aspx?PageID=cd316829-0381-452d-acfa-9da4f32cc17c

Erokhin, V. (2019). Tourism cluster development models in the Arctic: The opportunities for China to participate. In *Българо-китайски форум "Международни клъстерни политики" (Bulgarian-Chinese Forum "International Cluster Policies" Collection of Reports)*. Сборник доклади (pp. 19–41). Икономически университет-Варна (University of Economics - Varna).

Goffman, E. (1959). *Presentation of self in everyday life*. Garden City: Double Day Anchor Books.

Gössling, S., Hall, C. M., Ekström, F., Engeset, A. B., & Aall, C. (2012). Transition management: A tool for implementing sustainable tourism scenarios? *Journal of Sustainable Tourism, 20*(6), 899–916.

Gudlaugsson, T., & Magnússon, G. (2012). North Atlantic island destinations in tourists' minds. *International Journal of Culture, Tourism and Hospitality Research, 6*(2), 114–123.

Hook, L. (2019, September 20). Boom town: Greenland's climate change gold rush. *Financial Times*. Retrieved from https://www.ft.com/content/65dbd11a-d9a4-11e9-8f9b-77216ebe1f17.

Huang, M. F., Tang, C., & Weaver, D. B. (2017). The Arctic tourism experience from an evolving Chinese perspective. In Y. Lee, D. Weaver, & N. Prebensen (Eds.), *Arctic tourism experiences: Production, consumption and sustainability* (pp. 89–99). Wallingford, UK: CABI.

Hunt, C., & Stronza, A. (2014). Stage-based tourism models and resident attitudes towards tourism in an emerging destination in the developing world. *Journal of Sustainable Tourism*, *22*(2), 279–298.

Ioannidesv, D. (2019). Greenland's tourism policy making and the risk of overtourism. In C. Milano, J. Cheer, & M. Novelli (Eds.), *Overtourism: Excesses, discontents and measures in travel and tourism* (pp. 209–224). Wallingford, UK: CABI.

James, L., Olsen, L. S., & Karlsdóttir, A. (2020). Sustainability and cruise tourism in the Arctic: Stakeholder perspectives from Ísafjörður, Iceland and Qaqortoq, Greenland. *Journal of Sustainable Tourism*, *28*(9), 1425–1441.

Jørgensen, M. T., & Bertelsen, R. G. (2020). Chinese tourism in the Nordic Arctic–opportunities beyond the economic. *Scandinavian Journal of Hospitality and Tourism*, *20*(2), 166–177.

Kaae, B. C. (2002). Nature and tourism in Greenland. *Wilderness in the Circumpolar North: Searching for compatibility in ecological, traditional, and ecotourism values*. US Department of Agriculture, Forest Service, Rocky Mountain Research Station, Report RMRS-P-26, 43–53.

Kaiser, B., Huijbens, E., Aquino, J., Horbel, C., Reigner, N., Maher, P., … Turner, J. (2019). *Sustainable tourism development in the Nordic Arctic: Proceedings and Analysis*. London, UK: Polar Research and Policy Institute. Retrieved from https://polarconnection.org/wp-content/uploads/2019/06/Sustainable-Arctic-Tourism-Workshop-Report-15-June-2019.pdf.

Lojo, A. (2020). Young Chinese in Europe: Travel behavior and new trends based on evidence from Spain. *Tourism: An International Interdisciplinary Journal*, *68*(1), 7–20.

Lee, Y. (2020). Asia and Arctic tourism. *Scandinavian Journal of Hospitality and Tourism*, *20*(2), 105–109.

Lim, D. J., Ferguson, V. A., & Bishop, R. (2020). Chinese outbound tourism as an instrument of economic statecraft. *Journal of Contemporary China*, *29*(126), 1–18.

Martin, N., & Ren, C. (2020). Adventurous Arctic encounters? Exploring Chinese adventure tourism. *Scandinavian Journal of Hospitality and Tourism*, *20*(2), 126–143.

Ministry of Industry and Innovation. (2015). Road map for tourism in Iceland; Ministry of Industry and Innovation & The Icelandic Travel Industry Association. Retrieved from https://www.stjornarradid.is/media/atvinnuvegaraduneyti-media/media/Acrobat/Road-Map-for-Tourism-in-Iceland.pdf.

North Atlantic Tourism Association. (n.d.). *Tourism strategy 2015 – 2025*. Retrieved from https://www.northatlantic-islands.com/about-nata/strategy.html.

Nordic Council of Ministers. (2019). *Plan for Nordic tourism co-operation 2019 – 2023*. Retrieved from http://norden.diva-portal.org/smash/record.jsf?pid=diva2%3A1330360&dswid=-9768

Paik, W. (2020). The politics of Chinese tourism in South Korea: Political economy, state-society relations, and international security. *The Pacific Review*, *33*(2), 331–355.

Petridou, E., Olausson, P. M., & Ioannides, D. (2019). Nascent island tourism policy development in Greenland: A network perspective. *Island Studies Journal*, *14*(2), 227–244.

Plieninger, T., av Rana, H. Á., Fagerholm, N., Ellingsgaard, G. F., Magnussen, E., Raymond, C. M., … Verbrugge, L. N. (2018). Identifying and assessing the potential for conflict between landscape values and development preferences on the Faroe Islands. *Global Environmental Change*, *52*, 162–180.

Qu, H., Kim, L. H., & Im, H. H. (2011). A model of destination branding: Integrating the concepts of the branding and the destination image. *Tourism Management*, *32*, 465–476.

Rantala, O., Barre, S. D. L., Granås, B., Jóhannesson, G. Þ., Müller, D. K., Saarinen, J., … Niskala, M. (2019). *Arctic tourism in times of change: Seasonality*. Nordic Council of Ministers. Retrieved from https://norden.diva-portal.org/smash/get/diva2:1312957/FULLTEXT01.pdf.

Reigner, N. (2019). *Sustainable tourism management – Scoping report*. Recreation & Tourism Science. Retrieved from https://mfr.osf.io/render?url=https://osf.io/q9fv7/?direct%26mode=render%26action=download%26mode=render.

Reigner, N., Ren, C., Jóhannesson, G. Þ., Ásgeirss, M. H., & Árnsason, Ó. (2020). *Grant report: What lessons from Iceland's experience responding to tourism growth can Greenland use to prepare for tourism growth?* Recreation & Tourism Science. Retrieved from https://mfr.osf.io/render?url=https://osf.io/29m4d/?direct%26mode=render%26action=download%26mode=render.

Ren, C., Chimirri, D., & Abildgaard, M. S. (2020). *Toward culturally sensitive tourism: Report from Greenland. ARCTISEN – Promoting culturally sensitive tourism across the Arctic*. Lapland University Consortium, Multidimensional Tourism Institute. Retrieved from https://lauda.ulapland.fi/bitstream/handle/10024/64215/Arctisen_Greenland_report_FINAL.pdf?sequence=1.

Ren, C., Jóhannesson, G. T., Kramvig, B., Pashkevich, A., & Höckert, E. (2021). Twenty years of research on Arctic and Indigenous cultures in Nordic tourism: A review and future research agenda. *Scandinavian Journal of Hospitality and Tourism*, *21*(1), 111–121.

Rosen, J. (2016). Arctic dreams: As it pursues independence, Greenland seeks to develop its economy without ruining one of Earth's last pristine places. *Nature*, *532*(7599), 296–300.

Schionning, B. (2020, January 9). As the ice melts, Greenland considers its future. *BBC News*, *9*. Retrieved from https://www.bbc.com/news/business-51014148

Shackel, P. A. (2011). A snapshot of tourism in Greenland. *International Journal of Heritage Studies*, *17*(1), 81–88.

Smed, K. M. (2014). The culture of nature: Destination visitability in Ilulissat, Greenland. *Northern Review*, *38*, 269–293.

United Nations General Assembly. (2015). *International Year of Sustainable Tourism for Development, 2017*. Retrieved from https://www.unwto.org/archive/global/press-release/2015-12-07/united-nations-declares-2017-international-year-sustainable-tourism-develop.

Visit Faroe Islands. (2019a). *Ársfrágreiðing 2019*. Retrieved from https://issuu.com/visitfaroeislands/docs/vfi__rsfr_grei_ing_2019_single_web.

Visit Faroe Islands. (2019b). *Join the preservolution! A sustainable tourism development strategy for the Faroe Islands towards 2025*. Retrieved from https://www.visitfaroeislands.com/join-the-preservolution/strategy/.

Visit Greenland. (2018). *Tourism statistics report 2017 – Greenland*. Retrieved from http://www.tourismstat.gl/resources/reports/en/r15/Tourism%20Statistics%20Report%20Greenland%202017.pdf.

Visit Greenland. (2020). *Tourism statistics report 2019 – Greenland.* Retrieved from http://www.tourismstat.gl/resources/reports/en/r31/Tourism%20Statistics%20Report%20Greenland%202019.pdf.

Visit Greenland. (n.d.). *Towards more tourism, Visit Greenland tourism strategy 2020–2023.* Retrieved from https://traveltrade.visitgreenland.com/strategy/.

Volpe, M. (2020). The tortuous path of China's win-win strategy in Greenland. The Arctic Institute. Retrieved from https://www.thearcticinstitute.org/tortuous-path-china-win-win-strategy-greenland/#:~:text=To%20increase%20understanding%20of%20dimensions,the%20region%20(and%20beyond).

Zelenskaya, E. (2018). Geopolitics and tourism in the Arctic: The case of the national park 'Russian Arctic'. *Journal of Policy Research in Tourism, Leisure and Events, 10*(1), 33–47.

Part 3

Green cruise tourism and the potential new sea route between Asia and the Arctic

6 Renewable energy-based cruise tourism for Asians in the Arctic

A perspective from Northern Norway

Odin Foldvik Eikeland and Matteo Chiesa

Introduction

The Asian cruise market is booming. Currently representing 7% of global cruisers, the Asian cruise market is expected to expand exponentially and account for 20% of the global market by the end of this decade. This trend is especially driven by China, which has an increased number of middle-class consumers looking for new holiday experiences (Dowling & Mao, 2017). The Chinese cruise industry has changed significantly during the last decades. In 2007, there was almost no tourism activity, and there was no cruise ship (CS) traffic along the coastline of China (Mondou & Taunay, 2017). Seven years later, in 2014, there were 697,000 CS passengers leaving from harbours in China (CLIA, 2014). In 2018, the global CS market share shows that North America is the largest market, followed by Europe and Asia (Data Intelligence, 2020). By 2020, the Chinese central government, supported by the Ministry of Tourism guidelines, expects 4.5 million CS passengers and hopes to be the world's second largest cruise market after North America (Mondou & Taunay, 2017). The rapid growth of tourism activity in China and Asia, and the prospect for future growth, indicates that China and Asia will become major CS markets in the future. Still, so far in Norway, Asians only represents 1.8% (15,200 passengers) of the total numbers of cruise ship tourists in 2019, while Germans represent by far the largest market share (Innovasjon Norge, 2019).

In 2020, the COVID-19 pandemic hit the tourism industry heavily to the point of shutting down all CS tourism-related activities worldwide. In March 2020, the number of tourist arrivals fell by 57% globally, and the market will continue to be highly affected as long as the COVID-19 pandemic remains uncontrolled (World Tourism Organization, 2020). While the pandemic is still raging, it is not possible to predict how or if the CS market will remain the same as before 2020. It is also difficult to estimate which region will dominate future tourism growth. However, Singapore is attempting to start its tourism activities again, and as a consequence, CSs will begin sailing from November 2020. In order to keep the passengers and crew safe and minimise the risk of the COVID-19 disease, the CSs will make no stops and will return to the same port from which they departed (Penaloza, 2020). This indicates that some Asian countries are already trying to build up tourism

DOI: 10.4324/9781003039518-6

activities again, while the response to COVID-19 is still causing reduced activities in the rest of the world.

In addition to global growth, the Arctic region has become an increasingly popular tourist destination in recent years, partly due to (1) warmer temperatures and (2) an increasing demand for the so-called "last chance" tourism as a consequence of the impact of climate change (Palma et al., 2019). In some Arctic areas, CS tourism was one of the fastest growing economic sectors (before COVID-19). The pollution associated with this growth of tourism activity aggravates the local air quality and simultaneously increases greenhouse gas emissions (D'Aprile, 2018).

The increased awareness of the consequence of global climate changes in Arctic regions shows the need to reassess how the industry could minimise environmental impacts. As COVID-19 has shut down the industry temporarily, it is an opportunity for CS operators to develop strategies aimed at lowering the carbon footprint and thereby minimise local impacts in Arctic regions. This could be the beginning of a future CS market that makes the Arctic an increasingly attractive tourist region without compromising the environment.

Our purpose in this chapter is to (1) model the most popular CS harbour in Northern Norway and show that this is an excellent model to study how a solar energy system can be utilised to electrify cruise ship tourism in Arctic regions and (2) consider how such modelling-based renewable energy can be used to contribute to an environmentally sustainable cruise tourism by Asian visitors in Northern Norway.

A more sustainable Arctic tourism will have positive effects in terms of reducing local air pollution and greenhouse gas emissions, which is fundamental for the local population in CS destinations to positively experience the presence of tourists. In addition, developing a more environmentally sustainable tourism industry could have positive impacts on the experiences of tourists. For example, tourists have the opportunity to contribute to reducing emissions while at the same time experiencing the spectacular natural experience that the Arctic region offers.

Local context in Northern Norway

European ports have often been reluctant to install shore power facilities for CSs since the up-front costs are high and many ships do not have connection capabilities (Cruise Industry News, 2019). Consequently, ships generate electrical power by means of auxiliary engines on board while anchored in the harbour. This causes emissions of both greenhouse gases and conventional air pollutants. These have a negative impact on local air quality and contrast the "clean-nature" experience that the tourist industry is marketing in destinations throughout Norway. Studies performed by the sustainable transport group (Transport & Environment) state that the cruise ship operator Carnival Corporation emitted nearly 10 times more sulphur oxide in Europe than all 260 million cars in Europe in 2017 (Abbasov et al., 2019).

The future CS industry is expected to retrofit ships to have the ability to connect to shore power facilities. This could contribute to the electrification of CSs while harboured. When CSs are connected to onshore power sources, the load on the existing network increases rapidly. In worst-case scenarios, the existing network will not handle this load or meet demand. Consequences are voltage drops and power outages. The implication is that minimising the load on existing power grids becomes mandatory.

In Norway, more than 95% of power production comes from hydropower (International Hydropower Association, 2017). As a negative, installing hydropower stations requires major infrastructure developments and drastically impacts the local environment. This consideration arguably is sufficient to necessitate understanding the potential value of alternative energy technologies. Solar photovoltaics (PV) and wind power are so far the most prominent technologies for power generation from renewable sources (Creutzig et al., 2017; Gielen et al., 2019). However, there has been great opposition to wind power installations in Norway due to potential disturbance to wildlife and corresponding infrastructure development in otherwise uncontaminated areas. In this sense, solar energy systems mounted on buildings might easily gain support in the local society due to the perceived lower invasiveness of rooftop PV compared to wind parks.

Northern Norway is characterised by typical Arctic conditions with cold, long winters and polar nights between November and January where the sun is never above the horizon. The summer is short and cold, but it is characterised by long days with midnight sun between May and July, where the sun is never below the horizon. Northern Norway is well known for its beautiful nature experiences and is a popular tourist region where the tourism industry continues to increase yearly (Skinner, 2018). A significant part of tourist traffic comes via cruise ships, with most of the traffic arriving during summer (Willassen, 2019). In 2018, 330,211 people arrived on 283 ships, and in 2019, the number increased to 352,990 people on 356 ships (Kystdatahuset, 2020). In order to maintain its nature-friendly character, greenhouse gases and other pollutant emissions from harbours must be limited and ideally eliminated. The electrification of Arctic cruise tourism presents a promising pathway to achieving this reduction.

In Northern Norway, Tromsø and Honningsvåg are by far the most popular destinations (Kystdatahuset, 2020). Tromsø is the largest city in Northern Norway (Skinner, 2018). A significant part of tourist traffic comes via CSs, with most of the traffic arriving during summer (Willassen, 2019). In 2018, 142,348 people arrived aboard 119 large ships, and in 2019 the number increased to 155,160 people aboard 122 ships (Tromsø Havn, 2019). Honningsvåg, the second most popular CS destination in Northern Norway, is a popular destination due to the famous Nordkapp or North Cape plateau, which is often referred to as the northernmost location in Europe. The plateau attracts hundreds of thousands of tourists every year, with tourists from Asia accounting for the largest increase in most recent years (NTB, 2017).

The case of Tromsø in Northern Norway

Tromsø excels as the most popular tourist destination in Northern Norway in terms of the number of visiting cruise ships (Kystdatahuset, 2020). In this chapter, a specific case study was performed for this location to investigate the potential for utilising solar energy as the main source for electrifying the harbour to develop Tromsø as a more environmentally sustainable tourist destination.

Number of ships and visitors in Tromsø

The statistics on the number of ships and tourists arriving each year from 2013 to 2019 were provided from Kystdatahuset. These are provided in Table 6.1.

The decline in the number of CSs experienced between 2014 and 2016 is probably due to the aftermath of the 2014 oil price collapse, which resulted in slower economic growth (Prospects, 2018). From 2016 onwards, the number of CSs has increased together with the number of passengers. The data for the number of visiting CSs per month in the period ranging from 2013 to 2019 has been averaged and listed in Table 6.2.

This distribution shows how the great majority of CSs arrive during the summer tourist season, with a peak in June. The high number of visiting CSs in March is due to interest in northern lights tourism during winter (Skinner, 2018).

Potential for electrifying Tromsø harbour

In this section, the energy demand and solar availability are assessed to address the energy demand/generation match between cruise ships and solar energy facilities.

Energy demand

When cruise ships dock in harbour, they enter into "hotel mode". When in hotel mode, the ships require large amounts of energy to operate air conditioning, hot

Table 6.1 Statistics showing the number of CSs visiting Tromsø every year since 2013

	2013	*2014*	*2015*	*2016*	*2017*	*2018*	*2019*
No. of Ships	104	110	103	91	103	117	122
No. of passengers	108,681	111,631	111,190	102,495	125,455	141,945	155,160

Table 6.2 Averaged distribution of the numbers of ships visiting harbours in Northern Norway between 2013 and 2019

Month	*J*	*F*	*M*	*A*	*M*	*J*	*J*	*A*	*S*	*O*	*N*	*D*
CSs	0	3	8	1	6	36	31	14	3	2	2	1

water, lights, restaurants, and heating. These must be available all the time for the passengers and the crew (Graw & Faure, 2010; Simonsen, 2014). If a harbour does not offer shore power, the ships run auxiliary engines on board, releasing CO_2, SO_2, NO_x, and other pollutants into the atmosphere. In addition, the engines create noise pollution which disturbs residents in nearby areas.

Statistics regarding cruise ship traffic are provided by Tromsø harbour. The statistics provided show an average ship size has a maximum capacity of 1,516 passengers (Tromsø Havn, 2019). Information regarding typical power load has been provided upon request from the port of Kristiansand (Sikveland & Wangen, 2019), which is the only port in Norway that has shore power facilities to power CSs. Based on information regarding cruise ship traffic and typical power load, it is possible to calculate energy demand while a ship is at dock.

The average number of hours that ships dock in Tromsø varies by month from 25 hours in February to seven hours in September (Willassen, 2019). However, almost every CS that visits arrives between 07:00–09:00 in the morning. This suggests a potential good temporal matching between solar energy generation and energy load (Willassen, 2019). Based on a typical power load of 5 MW (Sikveland & Wangen, 2019) and the average number of docking hours, it is possible to estimate the aggregated monthly energy consumption using the following equation:

$$\text{Energy load}[\text{MWh}] = \text{No. boats} \times \text{Hours visited}[\text{h}] \times \text{Power of ship}[5\ \text{MW}]$$

Photovoltaic yield simulation

The energy yield from a typical photovoltaic (PV) system was simulated using PVsyst software (PVsyst, 2019). The simulation was performed using meteorological data from Meteonorm V7, which includes global horizontal irradiance (GHI), diffuse horizontal irradiance, wind speed, and temperature (PVsyst-Meteonorm, n.d.). On the basis of assuming the average power load from CSs to be 5 MW when in harbour, we have used a 6 MW_p PV system for the simulation. Such a system covers an area of approximately 32,736 m^2 and can be installed on the existing rooftops of industrial buildings in Tromsø harbour. This covers a total area of approximately 55,000 m^2.

PV power output and charging load from cruise ships

Since the potential for PV power output and CS traffic is highest during summer, the energy balance assessment is primarily of interest during the summer period between April and September. The monthly energy consumption and PV power output are given in Figure 6.1.

Figure 6.1 shows that for April, May, and September, the aggregated PV yield is sufficient to fully cover all the consumption from CS traffic. However, in June, July, and August, when the traffic is at its highest, the PV yield from the 6 MW_p system cannot cover the full demand.

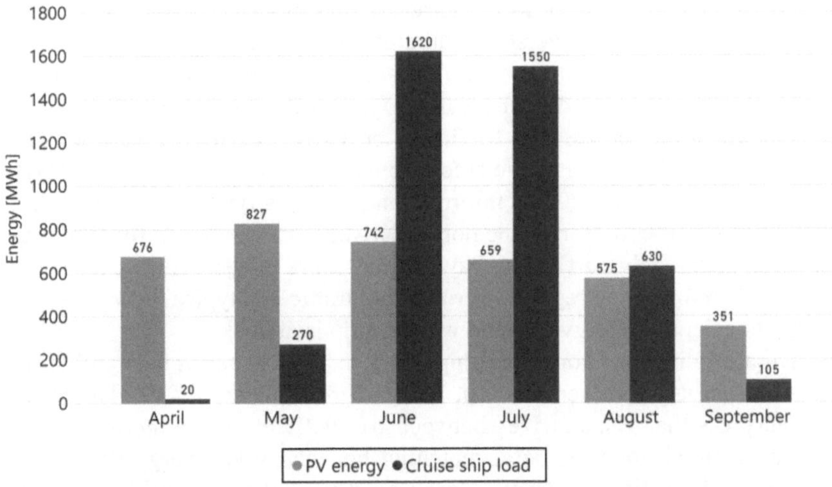

Figure 6.1 Energy balance between aggregated values during summer in Tromsø. The PV energy generation is indicated by the darker shade. The load from cruise ships in June-August is significantly higher than the generated PV energy

Economic considerations

So far, the benefits from an environmental standpoint of transitioning CSs from onboard power to renewable shore power have been investigated. However, in order to gain an understanding of what measures would be needed to implement this solution in practice, it is necessary to consider the cost structure of the proposed system in comparison to the current standard.

At present, CSs use their auxiliary engines as generators to provide onboard power while docked. These engines are fuelled, for the most part, by maritime fuel oil (MFO), a low-cost but highly polluting petroleum fuel. The lowest grades of fuel oil are banned in the Arctic for environmental reasons (International Maritime Organization (IMO), 2019). Marine Gasoil (MGO), a refined maritime fuel, is therefore assumed as the energy source for onboard power generation (Oiltanking, 2019). Global market data suggest a global average price for this fuel of ~$650 per metric tonne (Ship & Bunker, 2019). High-quality oil and new diesel engines on ships are recorded to have efficiencies up to 45% (Hundertmark, 2019). However, as the ships are assumed to use MGO for onboard electricity generation, and the engines on the ships are most likely not going to be new and/or high-quality engines, an efficiency in the range of 30–45% could be assumed as reasonable. To match a near future scenario, an efficiency of 33% was chosen. Based on this, we can calculate the US dollar cost of a business-as-usual scenario in which ships continue burning MGO for electricity.

Beginning from an energy content for MGO of 45 kJ/g = 12.500 kWh/tonne and assuming an efficiency of 33% for onboard electricity generation (Hundertmark, 2019), the fuel cost per kWh would be:

$$F = \frac{\$650}{0.33 \times 12.500\ \text{kWh}} \cong \mathbf{15¢/kWh}$$

The Brent price can oscillate rapidly due to several market events, which in turn will affect the onboard electricity cost considerably. However, in this chapter, the global market data suggestions are used to provide a rough estimate of fuel electricity cost for CSs. For comparison, we consider global standard costs of rooftop photovoltaic systems under the conditions modelled in this study. According to the International Renewable Energy Agency, total installation costs of rooftop photovoltaic systems fell below \$2/W across most of Europe in 2017 (International Renewable Energy Agency (IRENA), 2018). As both hardware prices and other costs continue to fall, this value now represents a conservative estimate for rooftop PV costs. On the other hand, there are vast differences in hardware prices between different countries, which in turn could lead to significantly higher prices than indicated by IEA. Therefore, the price of \$2/W is considered as a reasonable value for the PV system analysed in this chapter. Using this value, the energy yield simulation results given in the previous section, we can calculate the levelised cost of electricity (LCOE) from the rooftop system:

$$\text{LCOE} = \frac{\text{Total costs over lifetime}}{\text{Sum of total energy generated}} = \frac{C + \sum_{i=1}^{N} \frac{O_i}{(1+r)^i}}{\sum_{i=1}^{N} \frac{E_i}{(1+r)^i}}$$

The LCOE measures the average costs of electricity generation for a technology over its lifetime. The LCOE are given as the ratio between all costs during the lifetime and the total energy generated. For investment decisions, the costs should be low and the energy generation during its lifetime should be large. Therefore, the LCOE is a fundamental parameter for investment strategies when choosing between different technologies. The technology providing the lowest LCOE should be the one considered.

The parameters in equation (3) are: C, the installation costs including all hardware, labour, and soft costs; O_i the operating cost for year 1 in USD/kW_p/year; E_i is the energy produced in year 1 in kWh/kW_p/year; and r is the discount rate which is related to the cost of financing the project. The National Renewable Energy Laboratory (NREL) estimates suggest annual maintenance costs for rooftop PV systems at \$20/kW_p (Fu, Feldman, & Margolis, 2018); international experience in solar energy project finance has seen PV projects in countries with pro-solar policies financed heavily by loans with interest rates of 4% or less (Apostoleris & Chiesa, 2019). The simulation results on the 6MW system gave 4,795MWh of annual energy delivered to grid in the first year. Calculating the LCOE based on these values, and assuming a 1% annual escalation of operation and maintenance (O&M) costs and 0.5% degradation of PV output per year, we find that the cost of electricity from the PV system would be 22.5¢/kWh (in USD) over 20 years. If the system is properly maintained in order to extend its lifetime to 30 years, the LCOE would fall to 18.5¢/kWh.

However, given the prevailing low interest rates in Norway, a public investment programme could sustainably finance such a project at substantially lower rates. If the project were instead financed at a 2% rate of return, the LCOE over 30 years would fall to 15¢/kWh, matching the fuel cost of the business-as-usual scenario. Hence, the combination of investment by a public entity expecting low returns on its investment and revenue from cruise ship operators paying the same price for shore power as they would for fuel allows the PV project to break even over its lifetime.

Discussion

The study shows that the renewable-based Arctic CS tourism could be viable in a technological (the solar resources can cover most of the demand) and in an economical way. However, so far, the social impacts have not been discussed. In order to develop a green transitioning of the cruise ship industry, the social aspects are fundamental to consider. Therefore, it is necessary to consider how such transitioning affects the locals and all other actors involved. To create an emerging market, the investors and actors involved must see how this transitioning could be economically attractive. The study has shown that the proposed solution is economically viable compared to fossil fuel solutions. This shows that the proposed solution should be of interest among investors.

The transitioning should not impact the locals (inhabitants of Tromsø harbour) in a way that is perceived as negative. Moreover, electrification of CS harbours is beneficial as the cruise ships will not release local pollutants and create a local noise disturbance. This is of especial importance in Tromsø as the CS harbour is located in the middle of the centre where a lot of people are working and enjoying city life.

When using solar energy systems as the main power source for electrifying the harbour, the need for upgrading the existing electricity distribution network is not critical. Upgrading distribution networks has been a major source of conflict due to its huge environmental impact. In addition, solar energy systems provide the potential to utilise already existing infrastructure, as the system could be mounted on existing buildings with "free" space on roofs and façades. This is in contradiction to hydropower and wind power where the environmental impacts of such installations have led to major public opinion and unrest. Therefore, this chapter proposes a solution that has minimum impacts in terms of installation and development of existing infrastructure and has positive impacts as a power source with no local pollution in terms of noise disturbance and greenhouse gas emissions.

Asians are the fastest growing market in renewable energy installations and had the largest global share of renewable electricity capacity additions in 2007–2019 (IEA, 2020). In addition, Asians have shown dominance in green energy transitions and are now moving rapidly towards a non-fossil society (Korsnes, 2020).

In Tromsø harbour, a Polar tour operator company has started Svalbard sailings from Tromsø with a catamaran ship charged by solar energy systems installed on the roof of the ship, and the co-founder stated that "The tourists care about their green impact, so the business is obvious" (Nilsen, 2020). This indicates that industry actors have started a pathway towards a more emission-free Arctic tourism, and customers seem to be positive to such transitions.

The increased interest for green transitions gives a perfect opportunity to develop tourism that is more sustainable in terms of environmental (less pollution), economic (the technology cost could compete with existing solutions), and social (the tourists and locals will have a more positive experience) perspectives.

As a downside, if electrifying the harbour produces an uncontrollable increase in tourism and results in overtourism, there could be potential conflicts. The locals could feel that tourists are occupying their cities when the season is at its most hectic. This has been a challenge in other popular Arctic tourist destinations such as in Lofoten, where the locals struggle with overtourism (Becker, n.d.; Nikel, 2019) which generates negative impacts for the locals.

In addition, the more tourists visit, the more need to satisfy these visitors. The visitors often want to explore local shops where they can by "unique" things to bring home to their families. The shops require more infrastructure, which again further requires additional energy. Depending on the definition of sustainability, all such new developments are not entirely sustainable. Therefore, developing a more environmentally sustainable tourism industry could lead to less sustainable tourism destinations.

In summary, electrifying Arctic CS tourism has a positive outcome in terms of the environment (the local pollution and noise disturbance are reduced) and is in line with the global focus on green transitions. The tourists and locals seem to be positive to such transitions, so long as tourism is developed in a controlled way.

Conclusions

We have proposed a solution to utilise solar energy technologies to deal with a specific problem related to greenhouse gas emissions at Arctic tourist destinations, namely, increased cruise ship tourism in the Arctic. To this end, the solar energy potential for Tromsø (Norway) has been evaluated by simulating the PV yield from a 6 MW_P system by exploiting PVsyst. The total energy consumption from the CSs was estimated based on information about typical load, combined with statistics about the number and the types of ships visiting Tromsø during the year. The aggregated monthly PV yield and CS load profiles were compared during the summer tourist season between April and September. Our results have shown that for the specific PV system in this study, the aggregated CS load could be fully covered in April, May, and September. Unfortunately, it could not be entirely covered in June, July, and August when the traffic is largest.

To fully utilise the proposed PV system in this study and make fertile ground for a more environmentally sustainable Arctic tourism, we believe that a potential business model could consider both (1) local power production companies and governments and (2) the inhabitants to finance most of the system (each inhabitant could invest in a part of the system). The PV system would continuously interact with the existing distribution grid where excess energy can be sold—investors would thus make a profit. In this way, PV generation would be always utilised. In addition, to create a driving force for an emerging market for renewables and shore power systems, governments could require the cruise ships to connect to the facility when visiting. If ships were not willing to connect (due to high costs) or did not have the possibility to connect, they could be told not to dock in Tromsø harbour. A less aggressive potential solution would be to pay an additional price for releasing local pollution. This could be made significantly higher than using the shore power facility.

In summary, we have used the particular location of Tromsø as a model system to discuss the main environmental, technical, social, and economic aspects that should be taken into account when implementing renewables to cope with an increasing Arctic tourism. It is hoped that this work can provide a driving force for the further development of solar energy to support a more environmentally sustainable tourism in the Arctic region.

Acknowledgements

O.F.E and M.C acknowledge the support from the Arctic Centre for Sustainable Energy (ARC) at UiT-the Arctic University of Norway through Grant No. 310026.

References

Abbasov, F., Earl, T., Jeanne, N., Hemmings, B., Gilliam, L., & Calvo Ambel, C. (2019). *One corporation to pollute them all*. Brussels: European Federation for Transport and Environment AISBL.

Apostoleris, H., & Chiesa, M. (2019). The role of financing in achieving ultra-low solar electricity prices in the Middle East. IEEE Photovoltaic Specialists Conference 46th Photovoltaic Specialists Conference (PVSC), 16–21 June 2019, Chicago, IL, USA.

Becker, E. (n.d.). Arctic Crush – The friendly residents of one of the most remote parts of Norway have begun to organize to combat overtourism. *Travel Weekly*. Retrieved November 08, 2020, from https://www.travelweekly.com/Europe-Travel/Arctic-crush-Norway-combats-overtourism

CLIA. (2014). *Asia cruise trends, 2014: Analysis, assessment, appreciation*. Washington, DC: Cruise Lines International Association.

Creutzig, F., Agoston, P., Goldschmidt, J. C., Luderer, G., Nemet, G., & C.Pietzcker, R. (2017). The underestimated potential of solar energy to mitigate climate change. *Nature Energy*, *2*. doi:10.1038/nenergy.2017.140

Cruise Industry News. (2019, October 15). *Shorepower an ongoing debate*. Retrieved September 30, 2019, from https://www.cruiseindustrynews.com/cruise-news/19744-shorepower-an-ongoing-debate.html

D'Aprile, A. (2018, August 21). Arctic tourism: How the Great North is becoming the new exotic. *Future Earth*. Retrieved from https://www.climateforesight.eu/future-hearth/arctic-tourism-how-the-great-north-is-becoming-the-new-exotic/

Data Intelligence. (2020). Cruise tourism market, size, share, opportunities and forecast, 2020–2027. *Data Intellegence*. Retrieved from https://www.datamintelligence.com/research-report/cruise-tourism-market

Dowling, R., Weeden, C., & Mao, I. (2017). Cruising in Asia, with a focus on China. In *Cruise ship tourism*, 2nd edition. CAB International. doi: 10.1079/9781780646084.0562

Fu, R., Feldman, D., & Margolis, R. (2018). *U.S. solar photovoltaic system cost benchmarck: Q1 2018*. U.S Department of Energy, Golden, CO: National Renewable Energy Laboratory.

Gielen, D., Boshell, F., Saygin, D., D.Bazlian, M., Wagner, N., & Gorini, R. (2019). The role of renewable energy in the global energy transformation. *Energy Strategy Reviews*, *24*, 38–50.

Graw, R., & Faure, A. (2010). *Air pollution emission inventory, for 2008 tourism season, Klondike Gold Rush National Heritage Park*. Alaska: Alaska Department of Environmental Conservation.

Hundertmark, J. (2019). *Electrification of the Hurtigruten fleet*. HAMM: HAMM-LIPPSTADT, University of Applied Sciences.

Innovasjon Norge. (2019). *Cruiseturismen i Norge 2019*. Oslo: Epinion.

International Energy Agency. (2020, May). *Renewable energy market update – Outlook for 2020 and 2021*. International Energy Agency. Retrieved November from https://www.iea.org/reports/renewable-energy-market-update/2020-and-2021-forecast-overview#abstract

International Hydropower Association. (2017, May). Norway. IHA-International Hydropower Association. Retrieved from https://www.hydropower.org/country-profiles/norway

International Maritime Organization (IMO). (2019). *Shipping in polar waters- International code for ships operating in polar waters (Polar Code)*. Retrieved from http://www.imo.org/en/MediaCentre/HotTopics/polar/Pages/default.aspx

International Renewable Energy Agency (IRENA). (2018). *Renewable power generation costs in 2017*. Abu Dhabi: IRENA.

Korsnes, M. (2020). Creating a world after whose image? In *Wind and solar energy transition in China* (pp. 161–179). Milton Park, OX: Routledge.

Kystdatahuset. (2020, August 23). *Kystdatahuset-datavarehus*. Kystverket. Retrieved August 23, 2020, from https://kystdatahuset.no/

Mondou, V., Taunay, B., Dowling, R., & Weeden, C. (2017). Is China the new Goldmine for cruise companies? In R. Dowling & C. Weeden (Eds.), *Cruise ship tourism*, 2nd edition (pp. 546–561). Oxford: CABI.

Nikel, D. (2019, May 06). Nusfjord locals: Enough is enough. *Life in Norway*. Retrieved November 08, 2020, from https://www.lifeinnorway.net/nusfjord-locals/

Nilsen, T. (2020, February 4). Electric cruise pioneers start Svalbard sailings with catamaran to be charged by the sun. *The Barents Observer*. Retrieved from https://thebarentsobserver.com/en/travel/2020/02/electric-cruise-pioneers-sail-svalbard-catamaran-be-charged-sun?mc_cid=5def757a4d&mc_eid=a932aa2287

NTB. (2017, January 17). Turistrekord på Nordkapp. *ABC nyheter*. Retrieved from https://www.abcnyheter.no/reise/2017/01/19/195271687/turistrekord-pa-nordkapp

Oiltanking. (2019). *Important terms from A to Z*. Retrieved November 16, 2019, from https://www.oiltanking.com/en/news-info/glossary/details/term/marine-fuels.html

Palma, D., Varnajot, A., Dalen, K., Basaran, I., Brunette, C., Bystrowska, M., & Ronge, T. (2019). Cruising the marginal ice zone: Climate change and Arctic tourism. *Polar Geography*, *42*(4), 215–235. doi:10.1080/1088937X.2019.1648585

Penaloza, M. (2020, October 8). Singapore will launch cruises again, but with no destinations. National Public Radio (NPR). Retrieved from https://www.npr.org/sections/coronavirus-live-updates/2020/10/08/921636344/singapore-will-launch-cruises-again-but-with-no-destinations?t=1604768145581

Prospects, G. E. (2018). *Special focus 1. With the benefit of hindsight: The impact of the 2014–16 oil price collapse*. World Bank Group. doi:10.1596/978-1-4648-1163-0_Focus1

PVsyst. (2019). *PVSyst photovoltaic software*. PVSyst. Retrieved from https://www.pvsyst.com/

PVsyst- Meteonorm. (n.d.). *Meteonorm data and program*. Retrieved from https://www.pvsyst.com/help/meteo_source_meteonorm.htm

Ship & Bunker. (2019). *Rotterdam bunker prices*. Ship & Bunker. Retrieved from https://shipandbunker.com/prices/emea/nwe/nl-rtm-rotterdam#MGO

Sikveland, T., & Wangen, E. (2019, September 30). *Project manager and electrical manager at Port of Kristiansand*. Retrieved 06 04, 2019, from https://www.portofkristiansand.no/

Simonsen, M. (2014). *Cruise ship tourism – A LCA analysis*. Sogndal: Western Norway Research Institution.

Skinner, T. (2018, September 5). Is Norway's Tromso the new Reykjavik? *Financial Times*. Retrieved from https://www.ft.com/content/a4fa6e78-ac3d-11e8-8253-48106866cd8a

Tromsø Havn. (2019). *Om Tromsø Havn*. Retrieved from https://www.tromso.havn.no/om-oss/

Tromsø Havn. (2019). *Statistikk*. Retrieved from https://www.tromso.havn.no/om-oss/om-oss/statistikk/

World Tourism Organization. (2020). *Impact assessment of the COVID-19 outbreak on international tourism*. World Tourism Organization. Retrieved 2020, from https://www.unwto.org/impact-assessment-of-the-covid-19-outbreak-on-international-tourism

7 The Northeast Passage

Connecting China and the Arctic through cruise tourism?

Minna-Liina Ojala, Juulia Räikkönen, and Tuomas Kiiski

Introduction

Following the gradual abolishment of strict travel restrictions from the 1990s onwards and the setting of policies promoting outbound tourism, China has become the largest tourism source market in the world (Arlt, 2013; Johnson et al., 2020). Improved income levels of the middle class, the crumpling of knowledge barriers through digitalisation, and a growing appreciation of clean nature have bolstered the intense growth of Chinese domestic and outbound tourism (e.g. Cai et al., 2020; Lee, 2020). In particular, regional cruise tourism in Asia is experiencing a boost among the Chinese middle class (e.g., Dowling & Mao, 2017; Hung et al., 2020; Mondou & Taunay, 2017; Park et al., 2019). Additionally, neighbouring Russia rates among Chinese tourists' top new destinations (Luo et al., 2020), and a rapidly growing number of Chinese tourists are heading to the Arctic (e.g., Jørgensen & Bertelsen, 2020; Lee, 2020).

Studies by Pashkevich and Stjernström (2014) and Zelenskaya (2018) indicate a growing interest by Chinese tourists towards the Russian Arctic National Park. The Park can be accessed via the Barents Sea that is mostly ice-free all-year-round. Recently, climate change has facilitated a transformation in conditions that make also the Northern Sea Route (NSR) in the Russian territorial waters more relevant for commercial shipping (e.g., Wang et al., 2020). Along with these developments, a relatively small but increasingly attractive cruise tourism market has been forming there. Russia has intentions to significantly invest in tourism in its Arctic regions (Staalesen, 2020), and, simultaneously, China is interested and involved in improving the commercial use of the NSR (Miheeva, 2019). Furthermore, tourism is one of the "soft means" through which China is increasing its presence in the Arctic (Kobzeva, 2019), and the popularity of travelling along the new Silk Road, of which the NSR is part, is expected to grow (Luo et al., 2020). However, research and data concerning tourism and cruise tourism in Arctic Russia are scarce (Zelenskaya, 2018).

In this chapter, we aim to address this research gap by analysing the current state of cruise tourism and shipping in Russian Arctic waters and especially the NSR that is in the global limelight. We use statistical research data collected from NSR authorities and combine it with qualitative and quantitative data gathered

DOI: 10.4324/9781003039518-7

from tour operators' websites. The empirical results are discussed in light of Veblen's (1899) ideas on conspicuous consumption and Chinese tourists' preferences. We also take into account the post-COVID-19 development of cruise tourism. Overall, the chapter offers a stepping-stone for further research on the topical yet controversial issue related to the Arctic region's development as a unique tourism destination.

Arctic shipping and the NSR

The three main Arctic shipping routes (see Figure 7.1) that connect the Atlantic and the Pacific Oceans are the Northwest Passage, the Transpolar Sea Route, and the Northeast Passage (NEP). In this chapter, we concentrate on the Arctic NEP in the Russian territorial waters and especially on the NSR, an approximately 5,600 km long section of those waters. The shipping activity along the NSR is characterised by seasonality; the route is annually ice-free for about five months

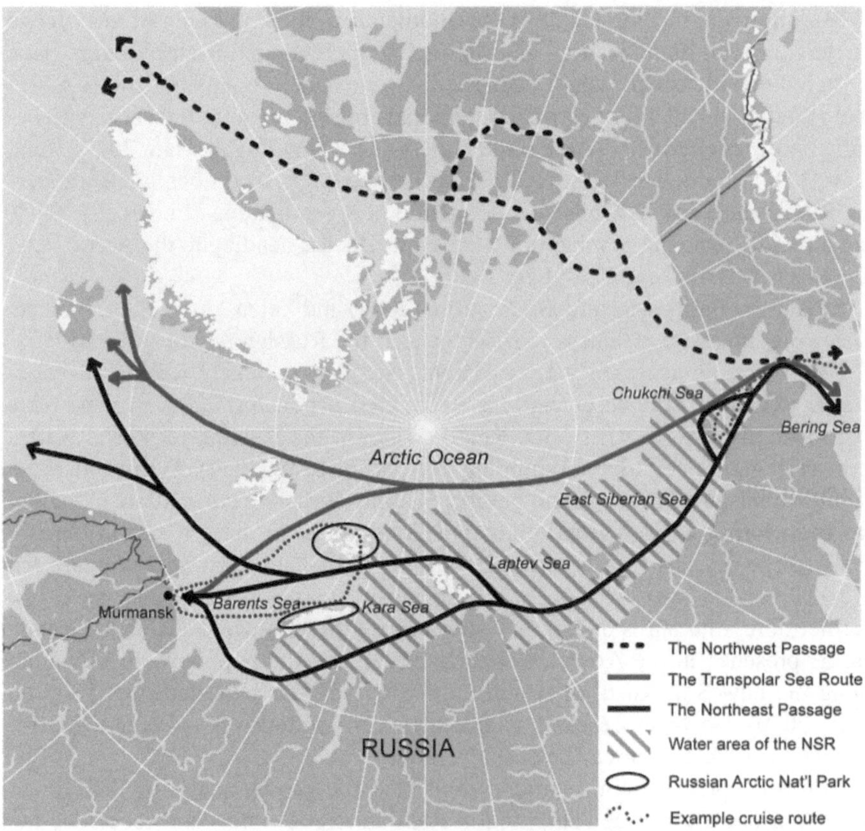

Figure 7.1 Arctic shipping routes and the approximate area of NSR (Map by A-M Kohijoki & M-L Ojala)

after which navigation requires a specialised fleet or/and use of icebreaker assistance (NSRIO, 2020). The navigation season has lengthened, and, currently, there is growing global economic interest towards the navigability of the NSR, as the Arctic represents a significant shortcut compared to conventional Southern shipping routes (e.g., Faury, 2019; Kauppila & Kiiski, 2020). In addition to potential time and fuel savings for shipowners, the NSR offers a rather secure operating environment, as Arctic waters have been absent from terrorism and piracy. Furthermore, the Arctic is significant for its oil, gas, and mineral resources, providing vast potential for economic development.

China's interest in the opening of Arctic sea routes has grown substantially due to the importance of effective logistics for global business competitiveness and securing energy and mineral resources (e.g., Kauppila & Kiiski, 2020). China has the world's largest shipbuilding industry, and as an increasingly powerful node of seaborne trade and energy flows, it is reshaping the future of global shipping. According to Wang et al. (2020), China could divert 20% of its oil shipping to the NSR. Thus, it is already interested and engaged in modernising the NSR's infrastructure and mechanisms for commercial use (Miheeva, 2019).

However, Arctic shipping has several economic and geographical constraints. Several studies argue that optimism regarding the Arctic route's vast economic potential is overstated, and global shipping along the NSR is likely to remain marginal for decades (e.g., Buixadé Farré et al., 2014; Kiiski et al., 2018). That being said, the shipping volume of NSR has steadily increased from over three MT (2012) to 33 MT (2020) (Kauppila & Kiiski, 2020; "Northern Sea Route Cargo" 2021). However, the performance of transit traffic is still negligible, and many obstacles must be tackled before large-scale shipping can become a reality (Kauppila & Kiiski, 2020). First, the harsh and unpredictable conditions increase shipping costs. Even though global warming melts sea ice, ships will still need to break through ice in winter, meet drifting ice in spring and fall, and face long periods of dark and extreme cold temperatures (e.g., Faury, 2019). Second, the availability of expensively built and operated ice-classed ships is limited, and icebreaker escorts are often needed (Solakivi et al., 2018). Third, revenue-making is affected by a thin cargo base, seasonality, reduced sailing speeds, and delays that are characteristic to the area (Kiiski et al., 2018). Thus, the business potential of the NSR remains contested, and shipping companies have difficulties in realising business undertakings there (Lasserre et al., 2016).

Arctic cruise tourism

The harsh and challenging nature and subsequent peripherality have always limited human activity in the Arctic, but the same factors are also essential in forming its allure as an exceptional travel destination (Lee et al., 2017). During the past decades, technological improvements and modern infrastructure have made the region and its tourism experiences more accessible (Têtu et al., 2019). Moreover, the media has presented the Arctic as a vanishing destination—often pictured with polar bears and astonishing landscapes—promoting last-chance-tourism, experiencing the Arctic before it is too late (Lemelin et al., 2010; Palma et al., 2019). Thus, tourism has become a significant reason for a human presence in the Arctic (Larsen & Fondahl, 2015).

Maher (2013, 2017) has calculated and estimated the annual number of tourists in various Arctic destinations. These two studies indicate that since early 2000, tourism has grown substantially in all geographical regions and in most Arctic destinations. In the North American Arctic (Alaska, Canada, and Greenland), the number of tourists has increased from circa 1.8 million (Maher, 2013) to 2.7 million tourists (Maher, 2017) while in the Scandinavian Arctic (Iceland, Norway, Sweden, and Finland), the increase was from circa 6.5 million (Maher, 2013) to 7.1 million tourists (Maher, 2017). Notably, obtaining reliable data on tourism statistics in the Arctic, and, especially the Russian Arctic, is challenging (Lee et al., 2017; Maher, 2017). According to rough estimations, the Russian Arctic attracted a few tens of thousands of tourists in the mid-2000s, but, in more recent estimations, the number has increased to approximately 500,000 annual tourists (Golubchikov et al., 2018; Tzekina, 2014).

Similarly, research and reliable statistics concerning Arctic cruise tourism are even scarcer. According to the calculations and estimations of Têtu et al. (2019), the most popular destination was Reykjavik with 96 vessels, yet the number of passengers was not available. Both Alaska and Greenland had almost 60 annual vessel visits, but Alaska had over a million passengers in 2018 while Greenland attracted about 25,000–30,000 annual passengers. Svalbard had 48 vessel visits and in 2016, about 70,000 passengers, and, finally, Arctic Canada attracted 14 vessels annually and about 3,500–6,000 passengers (Têtu et al., 2019). Again, information concerning Arctic cruise tourists in Russia is scarce, but Pashkevich and Stjernström (2014) and Zelenskaya (2018) reported that the number of annual cruise tourists visiting the Russian Arctic National Park in 2012–2016 ranged from less than 500 to some 1,200 annual visitors. For example, in 2016, nine cruise vessels visited the Park and carried 954 tourists from 40 countries. Here, Chinese tourists formed the largest cruise tourist segment, and that segment seems to be growing although the numbers are still modest: from circa 220 tourists in 2012 (Pashkevich & Stjernström, 2014) to circa 270 in 2016 (Zelenskaya, 2018). Although the Chinese cruise tourist market is currently moderate, the Chinese inbound tourist market to Russia in 2019 represented 8% of the total inbound tourist market (see chapter by Kobzeva in this volume). Subsequently, there is potential for growth in the cruise tourist market.

Chinese cruisers and conspicuous consumption

The sizeable Chinese middle class has altered the dynamics of global tourism as well as is reinvigorating the popularity of cruise tourism (Hung et al., 2019, 2020; Lee, 2020). In general, cruise ship tourism became the fastest-growing tourism sector in the 1980s (e.g. Clayton & Karagiannis, 2008). In 2009, there were about 17 million ocean cruise passengers worldwide, and by 2019, their amount almost doubled to 30 million (Cruise Lines International Association, 2020a). Nearly half of them come from the United States, but Chinese tourists form the second largest group (Cruise Lines International Association, 2019a). Their share grows rapidly, as the industry is booming in China (Dowling & Mao, 2017; Hung et al., 2019; Mondou & Taunay, 2017; Park et al., 2019). The number of cruise passengers from mainland China increased from 217,000 in 2012 to over 2 million passengers in 2018 (Cruise Lines International Association, 2019b). Although the COVID-19 pandemic halted international tourism in 2020, the pause is likely to remain temporary, especially if adjustments are made to guarantee the safety of the passengers (Renaud, 2020).

The most common cruise option for Chinese people is a domestic or regional cruise in Asia (Cruise Lines International Association, 2019b). Tourism is a rather new phenomenon in China (Arlt, 2013), and most of the cruise passengers are first timers (Dowling & Mao, 2017). The preferred cruise length is less than a week, and the average Chinese cruise passengers appreciate diverse shopping possibilities, luxurious yet affordable accommodation, aesthetics and high-end entertainment on-board, and various port tours with attractions (Dowling & Mao, 2017; Hung et al., 2020; Park et al., 2019). Moreover, Arctic tourism is increasing in popularity among the Chinese (Jørgensen & Bertelsen, 2020; Lee, 2020), and a fraction of them is also interested in Arctic cruises (Cruise Lines International Association, 2019b). As noted, the Chinese have become the largest group of tourists in the Russian Arctic National Park, and their number is growing (Pashkevich & Stjernström, 2014; Zelenskaya, 2018).

Having and displaying (international) travelling experience is important for many Chinese tourists (Artl, 2013). For these tourists, a cruising experience is linked to high social status, and showing off evidence of having been cruising is essential (Dowling & Mao, 2017; Mondou & Taunay, 2017). Veblen (1899) termed this kind of behaviour as conspicuous consumption. The latter enables people to signal their higher social status through excess consumption. Indeed, more and more people in China belong to this group, the so-called leisure class. Signalling consumables may vary from different products and services to immaterialities and aesthetic, intellectual, and ethical preferences that distinguish the consumer as a higher-ranking individual (Patsiaouras & Fitchett, 2012). Thus, leisure activities and lifestyles—such as cruise vacations and tourism—can be considered as instances of conspicuous consumption (Liu & Li, 2020).

Data

In this study, we use data from traffic statistics and tour operators' websites to shed light on the volume of Arctic cruise tourism in Russian territorial waters and shipping activity in the NSR. The first body of material consists of information about vessel movements in the water area of the NSR between 2013–2019. This data is compiled from the statistics of the Northern Sea Route Administration (NSRA) (http://nsra.ru/) and the Northern Sea Route Information Office (NSRIO) (https://Arctic-lio.com/). The technical vessel details were derived from the Russian Maritime Register of Shipping (RMRS) (https://rs-class.org/en/). Due to the absence of vessels compliant with the strict definition of a cruise vessel in the RMRS, we used vessel types that had commercial passenger transport capabilities.

Secondly, we gathered cruise information from English-language tour operators' websites. Data collection took place in June–July 2020. The websites were found by entering the search words "Arctic Russia cruise tour?" to Google search engine. Approximately 200 first hits were checked, and 14 tour operators were found for the analysis. Although they are likely the majority and most popular companies, some relevant ones may not have been reached this way. The website material concerns tourist cruises scheduled for 2020 and 2021 in Russian Arctic waters. Mainland river and coastal cruises were not taken into account. Information concerning the vessel name, passenger capacity, tour dates, route, itinerary, and price range of 71 tour products was collected. Attention was also given to the way they were advertised. Many of the tour products were actually sold for the same cruise vessel and dates but offered by several companies. In these cases, the different products do not add to the number of cruise trips or passengers. However, the different price information is taken into account.

NSR traffic and Russian Arctic cruise tourism in recent years

The fleet involved in passenger transportation in NSR and nearby water areas are presented in Table 7.1. It seems that during the seven-year observation period, the fleet composition has remained relatively consistent. Yet, increasing passenger capacity (from 1,234 to 1,689) and at the same time decreasing time spent on the route (from 44 to 33 days) is notable. This trend in the increasing passenger capacity combined with the decreasing time spent en route cannot be fully analysed due to insufficient cruise tourism data. However, it is reasonable to assume that the trend indicates an increasing number of pure passenger transport. Given the current lack of data, we utilised the operators' website information in an effort to gauge current Arctic cruise traffic. This is explained in the following section.

At the time of research, based on website data, in 2020 and 2021, altogether 12 vessels were scheduled for a tourist cruise in Russian territorial waters (see Table 7.2). The passenger capacity range of these vessels is 50–335, the average being 150. Thus, it is clear that the cruises do not belong to the mass cruise tourism industry. Per year, about 27 cruises are provided. During the two years, 13 of them traversed the NEP, 25 focused on the Barents Sea (and North Pole), and 15 only cruised Far East Russia. The season starts in mid-June and lasts until the end of September, but most of the cruises are offered in July and August. The cruises last 13–27 days, the NEP cruise being the longest. However, some of these days are also spent outside Russian borders in the Bering Sea or Svalbard, for instance.

Excluding cruises from Murmansk to the North Pole, the calculated daily ticket price ranges between 530 and 3,840 USD depending on the vessel and the chosen cabin class, the average lowest ticket price being 900 USD per day and the highest 2,100 USD. The route and its navigability seem to affect the price levels greatly. For instance, *Spirit of Enderby*, with the cheapest prices, sails a considerable period in Bering Sea while trips to North Pole are the most expensive.

Table 7.1 Traffic volume in the NSR 2013–2019

The traffic volume of NSR between 2013 and 2019

Source: traffic data obtained from NSRA (2020), NSRIO (2020) and technical details from RMRS (2020)

Year	*NSR cargo volume by traffic type (million tonnes)*		*Total passenger carrying capacity*	*Average days spent on NSR*	*The annual share of passenger shipping[2] of the total days ships spent on NSR by ship type*			*Number of unique vessels in each passenger shipping[2] oriented ship type*		
	Total	*Transit[1]*			*Research*	*Passenger*	*Total*	*Research*	*Passenger*	*Total*
2019	31,5	0,7	1,689	33	8,9%	1,3%	**10,2%**	19	8	***27***
2018	19,7	0,5	816	32	8,5%	0,7%	**9,2%**	20	2	***22***
2017	10,7	0,2	1,003	41	11,7%	0,7%	**12,4%**	22	6	***28***
2016	7,5	0,2	1,306	35	12,5%	1,0%	**13,5%**	29	5	***34***
2015	5,4	0,04	1,136	29	12,0%	1,5%	**13,5%**	24	4	***28***
2014	4	0,3	1,573	34	15,1%	2,9%	**18,0%**	29	6	***35***
2013	4,2	1,4	1,234	44	16,9%	1,8%	**18,7%**	21	5	***26***

[1] Transcontinental passage between the Pacific and Atlantic Oceans, or vice versa

[2] In the absence of vessels compliant with the strict definition of a cruise vessel, we used vessel types that have commercial passenger transport capabilities

Table 7.2 Tourist cruises in the Russian Arctic waters 2020–2021

Tour operator data: annual cruises in the Russian Arctic waters (12 vessels) Source: 14 tour operators' websites, retrieved in June–July 2020

Vessel name[1]	*Number of trips*		*Average trip length (d)*	*Passenger capacity*	*Daily ticket price range*[3] *(USD)*
	2020	*2021*			
Akademik Shokalskiy	2	5[2]	25	50	533–1,150
Greg Mortimer	1	2	15	120	587–1,527
L'Austral	0	1	25	264	835–2,945
Le Boreal	1	4	21	232	769–3,077
MS Spitsbergen	2	0	15	335	771–2,281
National Geographic Endurance	2	3	25	126	1,112–2,804
National Geographic Orion	3	1	15	102	1,272–2,572
Ocean Adventurer	2	2	19	129	545–1,600
Sea Spirit	1	4	14	114	685–1,761
Silver Explorer	1	0	27	144	1,147–3,837
Spirit of Enderby (Professor Khromov)	2	4[2]	15	50	533–920
50 Years of Victory *(to North Pole)*	4	5	14	128	2,194–6,714
Total	**21**	**31**	–	–	-

[1] 50 Years of Victory is a nuclear powered icebreaker; other vessels are classified as passenger ships
[2] Uncertain information whether some of these trips are sailed with only either one of these vessels
[3] Without additional fees

In June 2020, most of the tours for 2020 and 2021 were almost fully booked. If all tickets were indeed claimed, these vessels would bring close to 3,300 annual tourists to the region. However, since the data was retrieved in summer 2020 and during the COVID-19 pandemic, some cruises might have been already cancelled and the information on some previous tours that year deleted. It is also likely that the 14 tour operators reached are not the only ones providing cruises in the area. Therefore, the numbers here should be deemed as an indication only of the current volume.

Based on the website data, annually 14 tourist cruises sail in the NSR water area. The traffic statistics show that the traffic volume of NSR has a constant year-by-year (2013–2019) increase, but the growth has side-lined passenger shipping (Table 7.1). Measured in total days spent en route per annum, the share of passenger shipping-related ship types (i.e. research and passenger vessels—that both may be in tourism use) has decreased from over 18% in 2013 to just 10% in 2019. This underlines the fact that, currently, the majority of vessel movements in NSR relate to either transportation of natural resources or supporting activities, e.g. supply of production sites. The launch of large gas and oil plants is also a likely explanation for the significant cargo volume growth in 2017, 2018, and 2019.

Arctic Russia for Chinese cruise tourists

As noted, Arctic cruise tourism is a niche market in the global cruise tourism industry, and cruise tourism in the Russian Arctic is a smaller sub-niche of this niche. The cruises are expensive and promoted as exclusive and extraordinary once-in-a-lifetime experiences, which is likely to increase their attractiveness and status signal value for conspicuous consumption. The uncertain, rugged, and pristine nature, isolation from the rest of the society, and the uniqueness of the experience are the selling points of Arctic Russian cruises. The common Chinese cruise passengers, however, appreciate diverse shopping possibilities, affordable luxury cruises, as well as various port tours during a voyage (Dowling & Mao, 2017; Hung et al., 2020; Park et al., 2019).

In general, Arctic cruises, especially in Russia, do not currently match these expectations. Although there are luxury cruises along the NEP, most of the current supply comprises of vessels originally built in the 1970s and 1980s. The vessels are rather small to be able to reach the difficult destinations; on-board services and cabins are often minimal and furnishings basic. To please the majority of Chinese cruisers, upgrades are required, as no glamorous services or entertainment are provided on most tours. Whereas the popular length of Asian cruises is less than a week, the Arctic tours are long without many possibilities to hop on and off. Mondou and Taunay (2017) note that this preference is due to the short length of holidays and that even the short near-abroad cruises are available only for the richest Chinese middle class.

Besides the wealthy, in their current form, Arctic cruises in Russia may suit another group of Chinese tourists—the adventurous ones. Indeed, according to Cai et al. (2020) and Martin and Ren (2020), Asians travelling to the Arctic do not represent conventional mass tourists. Instead, they seek novel experiences, unique landscapes, as well as physical and mental challenges to build new capacities and learn. These experiential purchases are instances of conspicuous consumption, too, as they are linked with identity demonstration (Bronner & Hoog, 2018). The sought-after challenges for learning may have some deeper link to Confucian *Chi* or wisdom as well (see Chapter 2 in this volume for more explanation). Hung et al. (2020) also note that for Chinese cruisers, the experience of escapism adds to luxury value creation—perhaps the isolation and uniqueness of the Arctic cruises could compensate for the lack of glamour. In this case, cruises may attract and blend people who are simultaneously engaged in dissenting conspicuous consumption.

With respect to infrastructure and geology, cruises in the Russian Arctic seem similar to the Canadian Arctic where, according to Têtu et al. (2019), the number of passengers is low due to a lack of supporting infrastructure that limits the market to smaller expedition-style vessels. In Russia, the uncertain political atmosphere might limit demand further, and bureaucratic difficulties such as obtaining a visa have hindered foreigners from arriving there (Zelenskaya, 2018). One recent limiter to cruising has been the COVID-19 pandemic that paused international tourism in 2020. However, it is suggested that in the post-COVID era, niche tourism in more remote destinations—such as the Russian Arctic—will gain popularity while mass tourism destinations might be avoided (Renaud, 2020). At the time of writing this chapter, new COVID-19 policies are being formulated for the cruise industry (Cruise Lines International Association, 2020b). If the health of passengers is safeguarded, tours on small vessels such as the ones operating in the Russian Arctic may become more attractive for a larger number of tourists. The 2020–2021 cruises were almost fully booked, which indicates high demand, and, based on recent studies, Chinese people are increasingly interested in travelling to Russia and cruising there. From 2021 onwards, various nationalities—Chinese included—may easily apply for an e-visa (Kubny, 2020). Further, Russia's aim to triple Arctic tourism by 2035 implies huge investments and boosts for the cruise sector (Staalesen, 2020). If the cruise supply is multiplied and upgraded, and as global warming continues to open the waterways, these changes are likely to increase commercial cruises in the Russian Arctic, including along the NSR.

Conclusion

This chapter focused on analysing the current state of cruise tourism in Russian Arctic waters with state-of-the-art traffic statistics concerning the NSR and information from tour operators' websites. The potential of Russian Arctic cruise tourism to accommodate the growing number of Chinese tourists was considered and discussed in light of Veblen's conspicuous consumption theory.

Based on the information retrieved from the websites, we estimate the annual number of cruise tourists to be close to 3,300 in the Russian Arctic waters. At least 12 vessels are used for tourism purposes, and, on average, they make two 20-day tours annually with the average passenger capacity of 150. Whereas Arctic cruises are a niche in the global cruise tourism market, cruises in the Russian Arctic are a sub-niche of this niche. In the NSR, the majority of vessel movements relate to the transportation of natural resources and supporting activities, and the share of passenger-related ship types has decreased from 18% to 10% between 2013 and 2019. However, increasing passenger capacity and decreasing time spent en route reflect an increasing role of pure passenger transport-related activity. Estimating the size of the passenger shipping market is notably difficult because of data availability limitations. Further research should be undertaken regarding the actual number of paying passengers, for instance, by contacting the tour companies as well as reckoning the deck plans and distribution of different price classes.

The current supply in Russian waters is unlikely to meet the expectations of most Chinese cruisers: the tours are long and expensive, and the current supply, with few exceptions, lacks the preferred luxury. "Chinese tourists" are, however, a vast and diverse group. A growing number are already travelling to Russia (e.g., Luo et al., 2020) and engaged in cruising there (Pashkevich & Stjernström, 2014; Zelanskaya, 2018). Russian Arctic cruises suit the wealthy and those interested in the Arctic who embrace adventurous experiences and unique landscapes. The expensive, extraordinary, and exclusive cruises are apt examples of conspicuous consumables, which is likely to increase their popularity, even among different types of people. Russia's—and China's—planned investments to Arctic tourism and the NSR are likely to diversify and multiply the offerings and the clientele alike. Furthermore, if niche tourism in remote destinations, such as the Russian Arctic, gains popularity in the post-COVID-19 era as envisioned (Renaud, 2020), these cruises are likely to attract even more customers and investments. However, increasing human activity in the Arctic causes damage directly to local environments and indirectly by contributing to climate change (e.g., Palma et al., 2019; Renaud, 2020; Wang et al., 2020). In this chapter, the focus was on estimating cruise tourism volumes in the Russian Arctic. Further studies are needed to assess Russian cruise tourism's environmental and social impacts.

References

Arlt, W. G. (2013). The second wave of Chinese outbound tourism. *Tourism Planning & Development, 10*(2), 126–133.

Bronner, F., & de Hoog, R. (2018). Conspicuous consumption and the rising importance of experiential purchases. *International Journal of Market Research, 60*(1), 88–103.

Buixadé Farré, A., Stephenson, S. R., Chen, L., Czub, M., Dai, Y., Demchev, D., … Kivekäs, N. (2014). Commercial Arctic shipping through the Northeast Passage: Routes, resources, governance, technology, and infrastructure. *Polar Geography, 37*(4), 298–324.

Cai, Y., Ma, J., & Lee, Y-S. (2020). How do Chinese travellers experience the Arctic? Insights from a hedonic and eudaimonic perspective. *Scandinavian Journal of Hospitality and Tourism, 20*(2), 144–165.

Clayton, A., & Karagiannis, N. (2008). Sustainable tourism in the Caribbean: Alternative policy considerations. *Tourism and Hospitality Planning & Development, 5*(3), 185–201.

Cruise Lines International Association. (2019a). *Leading passenger source countries in the global cruise industry in 2018*. Retrieved from www.statista.com/statistics/287114/cruise-passenger-share-by-source-country/

Cruise Lines International Association. (2019b). *2018 Asia cruise industry ocean source market report*. Retrieved from https://cruising.org/news-and-research/research/2019/july/2018-asia-cruise-industry-ocean-source-market-report

Cruise Lines International Association. (2020a). *Number of ocean cruise passengers worldwide 2009–2020*. Retrieved from www.statista.com/statistics/385445/number-of-passengers-of-the-cruise-industry-worldwide/

Cruise Lines International Association. (2020b). *Cruise industry COVID-19 facts and resources*. Retrieved from https://cruising.org/en-gb/ cruise-industry-covid-19-facts-and-resources

Dowling, R. & Mao, I. (2017). Cruising in Asia, with a focus on China. In R. Dowling & C. Weeden (Eds.), *Cruise ship tourism* (2nd ed.) (pp. 562–572). Wallingford: CABI.

Faury, O. (2019). Introduction. In F. Lasserre & O. Faury (Eds.), *Arctic shipping: Climate change, commercial traffic and port development* [E-reader version] (pp. 1–3). doi:10.4324/9781351037464

Golubchikov, Y. N., Kruzhalin V. I., & Nikanorova, A. D. (2018). Arctic tourism: State and prospects for Russia. *Geography, Environment, Sustainability, 11*(4), 5–13.

Hung, K., Huang, H., & Lyu, J. (2020). The means and ends of luxury value creation in cruise tourism: The case of Chinese tourists. *Journal of Hospitality and Tourism Management, 44*, 143–151.

Hung, K., Wang, S., Guillet, B. D., & Liu, Z. (2019). An overview of cruise tourism research through comparison of cruise studies published in English and Chinese. *International Journal of Hospitality Management*, *77*, 207–216.

Information & Analytical Agency PortNews, IAA PortNews. (2021, January 11). Northern Sea Route cargo traffic rose by 4.7% in 2020. Retrieved from https://en.portnews.ru/news/307248/

Johnson, P. C., Xu, H., & Arlt, W. G. (2020). Outbound Chinese tourism: Looking back and looking forward. *Journal of Policy Research in Tourism, Leisure and Events*, *12*(1), 1–7.

Jørgensen, M. T., & Bertelsen, R. G. (2020). Chinese tourism in the Nordic Arctic – Opportunities beyond the economic. *Scandinavian Journal of Hospitality and Tourism*, *20*(2), 166–177.

Kauppila, L., & Kiiski, T. (2020). The red dragon in global waters: The making of the Polar Silk Road. In E. Pongrácz, V. Pavlov, & N. Hänninen (Eds.), *Arctic marine sustainability* (pp. 465–485). Cham, Switzerland: Springer.

Kiiski, T., Solakivi, T., Töyli, J., & Ojala, L. (2018). Long-term dynamics of shipping and icebreaker capacity along the Northern Sea Route. *Maritime Economics and Logistics*, *20*, 375–399.

Kobzeva, M. (2019). China's Arctic policy: Present and future. *The Polar Journal*, *9*(1), 94–112.

Kubny, H. (2020, October 15). E-visa simplifies entry to Russia. *Polar Journal*. Retrieved from https://polarjournal.ch/en/2020/10/15/e-visa-simplifies-entry-to-russia/

Larsen, J. N., & Fondahl, G. (2015). *The second Arctic human development report (AHDR-II). Key policy relevant conclusions and suggestions for moving forward.* Retrieved from http://sdwg.org/wp-content/uploads/2015/05/NCR-7467355-v1-SDWG_AHDRII_BROCHURE_FINAL.pdf

Lasserre, F., Beveridge, L., Fournier, M., Têtu, P. L., & Huang, L. (2016). Polar seaways? Maritime transport in the Arctic: An analysis of shipowners' intentions II. *Journal of Transport Geography*, *57*, 105–114.

Lee, Y-S. (2020). Asia and Arctic tourism. *Scandinavian Journal of Hospitality and Tourism*, *20*(2), 105–109.

Lee, Y-S., Weaver, D. B., & Prebensen, N. K. (2017). Arctic destinations and attractions as evolving peripheral settings for the production and consumption of peak tourism experiences. In Y-S. Lee, D. B. Weaver & N. K. Prebensen (Eds.), *Arctic tourism experiences: Production, consumption and sustainability* [E-reader version] (pp. 1–8). doi:10.1079/9781780648620.0191

Lemelin, H., Dawson, J., Stewart, E. J., Maher, P., & Lueck, M. (2010). Last-chance tourism: The boom, doom, and gloom of visiting vanishing destinations. *Current Issues in Tourism*, *13*(5), 477–493.

Liu, T., & Li, M. (2020). Leisure & travel as class signifier: Distinction practices of China's new rich. *Tourism Management Perspectives*, *33*, 100627.

Luo, Q., Zeng, S., Qin, X., & Li, X. (2020). Travel along the new Silk Road: Understanding new trends in Chinese outbound tourism. In H. Pechlaner, G. Erschbamer, H. Thees, & M. Gruber (Eds.), *China and the New Silk Road* (pp. 187–195). Cham, Switzerland: Springer.

Maher, P. T. (2013). Looking back, venturing forward: Challenges for academia, community and industry in polar tourism research. In D. K. Müller, L. Lundmark, & R. H. Lemelin (Eds.), *New issues in polar tourism: Communities, environments, politics* (pp. 19–36). Amsterdam, Netherlands: Springer.

Maher, P. T. (2017). Tourism futures in the Arctic. In K. Latola & H. Savela (Eds.), *The Interconnected Arctic – UArctic Congress 2016* (pp. 213–220). doi:10.1007/978-3-319-57532-2_22

Martin, N., & Ren, C. (2020). Adventurous Arctic encounters? Exploring Chinese adventure tourism. *Scandinavian Journal of Hospitality and Tourism*, *20*(2), 126–143.

Miheeva, N. (2019). Chinese-Russian cooperation in the Arctic: Geopolitics and economics. *Earth and Environmental Science*, *302*, 012057.

Mondou, V., & Taunay, B. (2017). Is China a new goldmine for cruise companies? In R. Dowling & C. Weeden (Eds.), *Cruise ship tourism* (2nd ed.) (pp. 546–561). Wallingford: CABI.

Northern Sea Route Administration, NSRIO. (2020). Ice concentration. Retrieved, December 3, 2020 from https://arctic-lio.com/nsr_ice/

Palma, D., Varnajot, A., Dalenc, K., Basaran, I. K., Brunette, C., Bystrowska, M., … Ronge, T. A. (2019). Cruising the marginal ice zone: Climate change and Arctic tourism. *Polar Geography*, *42*(4), 215–235.

Park, S., Lee, W. S., Moon, J., & Heo, J. (2019). Examination of Chinese cruise tourists' attributes using a choice experiment. *Sustainability*, *11*, 3621. doi:10.3390/su1113362.

Pashkevich, A., & Stjernström, O. (2014). Making Russian Arctic accessible for tourists: Analysis of the institutional barriers. *Polar Geography*, *37*(2), 137–156.

Patsiaouras, G., & Fitchett, J. A. (2012). The evolution of conspicuous consumption. *Journal of Historical Research in Marketing*, *4*(1), 154–176.

Renaud, L. (2020). Reconsidering global mobility – Distancing from mass cruise tourism in the aftermath of COVID-19. *Tourism Geographies, 22*(3), 679–689. doi:10.1080/14616688.2020.1762116

Solakivi, T., Kiiski, T., & Ojala, L. (2018). The impact of ice class on the economics of wet and dry bulk shipping in the Arctic waters. *Maritime Policy & Management*, *45*(4), 530–542.

Staalesen, A. (2020, June 15). Moscow aims for a massive post-COVID boost in Arctic tourism. *The Barents Observer*. Retrieved from https://thebarentsobserver.com/en/travel/2020/06/Moscow-aims-post-covid-200-boost-arctic-tourism

Têtu, P. L., Lasserre, F., & Dawson, J. (2019). The evolution and relative competitiveness of global Arctic cruise tourism destinations. In F. Lasserre & O. Faury (Eds.), *Arctic shipping. Climate change, commercial traffic and port development* [E-reader version] (pp. 94–114). doi:10.4324/9781351037464

Tzekina, M. (2014). *Estimation of tourism potential of Russian Far North*. Moscow: Moscow State University.

Veblen, T. (1899). *The theory of the leisure class*. New York: Macmillan.

Wang, D., Ding, R., Gong, Y., Wang, R., Wang, J., & Huang, X. (2020). Feasibility of the Northern Sea Route for oil shipping from the economic and environmental perspective and its influence on China's oil imports. *Marine Policy*, *118*, 104006.

Zelenskaya, E. (2018). Geopolitics and tourism in the Arctic: The case of the national park Russian Arctic. *Journal of Policy Research in Tourism, Leisure and Events*, *10*(1), 33–47.

Part 4

Ethnicity, culture, experiences, and urbanisation

8 "When I hear there's going to be an Asian tourist …"

The role of Asian ethnicity in the planning process of guided tours in Svalbard

Eva Kotašková and Matouš Jelínek

Introduction

Asian tourists have become one of the dominant driving forces of the tourism industry. This is becoming increasingly significant in Svalbard, where the number of Asian tourists is currently rising. At the same time, Asians are perceived as rather specific types of tourists with particular expectations related to their ethnic origin.

When it comes to links between tourism and ethnicity, most research focuses on how ethnicity is constructed in "host" societies within so-called "ethnic tourism" where ethnicity becomes a commodity in a destination's tourism industry (Bott, 2018; Bruner & Kirshenblatt-Gimblett, 1994; Cohen, 2016; De la Maza, 2018; Drew, 2011 and others). As Wood has noted, "the advent and development of tourism may not only affect which ethnic identities are asserted, but which ethnic markers are chosen to symbolize [*sic*] group membership and culture" (Wood, 1998, p. 222). This applies also to ethnicity within the "guest" societies; the tourists, herein, might be described by the concept of an "Asian tourist". Constructing ethnicity from this perspective is rather understudied, and, generally, research on tourists and tourists' behaviours focus on nationality, gender, or class rather than ethnicity (for an overview, see e.g. Özdemir & Yolal, 2017). At the same time, most Asian tourism research, as well as common sense understandings, work with the concept of Asianness rather unproblematically, reducing Asianness primarily to East and South Asia (Hall, 2009). Since Asia is the world's largest and most populated continent and consists of people of numerous cultures, languages, and origins, such a lack of reflection has both practical and academical consequences (Hall, 2009)—e.g. it silences the diversities among Asians, especially, the particularities of variables creating everyday life situations and social structures.

In this chapter, we aim to explore the role of the Asian ethnicity of tourists in the planning process of guided tours. Further, we examine what the common labels and expectations of a typical Asian tourist are, and we aim to elaborate on the role that guides' interpretations play in the tour planning process. We conceptualise Asianness as an ethnic category and present a constructivist approach to ethnicity. We interpret ethnicity as a contextual, situational, and relational concept, and we show how this emerges in the process of planning guided tours.

DOI: 10.4324/9781003039518-8

During this process, the images and expectations associated with Asianness play a significant role as a source of information. These images and expectations help guides plan tours and adjust them according to individual as well as group skills and abilities. We show that the role of Asianness plays a significant role in the initial phases of the planning process when expectations and interpretation of experiences are the main source of information for guides. Further, during the planning process, the significance of Asianness as an ethnic category decreases when guides replace initial expectations with direct observations of tourists' abilities as the main source of information.

Conceptualising Asianness

From the logic of the term, "Asian" is an attribute of a person coming from, or living in, Asia. The category Asian refers to a combination of regional and phenotypic traits, such as skin colour, facial features etc. (Anthias, 1992, p. 423). In everyday life, people with those attributes are associated with certain images and expectations of characteristic behaviours. This is an elementary mechanism of ethnicity as a concept (see Eriksen, 2010; Fenton, 1999; and others). It means that we all probably have an idea who is and who is not Asian and what their typical characteristics and behaviour are. Readers should note that in this chapter, we use "they" as a nonbinary pronoun to follow gender-balanced language principles.

Obviously, a term including thousands of ethnic groups speaking thousands of languages is too general to be unproblematically used in academic writing. As Lee and Ramakrishnan (2019) suggest, the term "Asian" is confusing, unclear, and problematic because of its lack of consistent usage—people in daily life as well as in academic texts or legal definitions often associate the category Asian only with specific parts of Asia, typically with Southeast Asia (Chinese, Japanese, and Koreans), and exclude others (typically Indians, Filipinos, Arabs, or Russians) (see Bhopal, Phillimore & Kohli, 1991; Hall, 2009; Ocampo, 2016; and others). This is the reason why we need to explain to whom we are referring when we talk about Asians.

In this text, we conceptualise the term Asian as an ethnic category. The concept of ethnicity may be traced to the 19th century, when it was created as a way to categorise cultural differences within the context of modernity and its changes in the conceptualisation of social order through the lens of nationalism (see Eriksen, 2010 or Tonkin, McDonald & Chapman, 1989). Such concepts as ethnicity, race, and nationalism are kindred terms that refer to the idea of a common origin of members of social collectivities (Anthias, 1992).

There are two theoretical perspectives of the historical aspects of nation and ethnicity: primordialist and constructivist. From a primordialist perspective, ethnic and national origin is "a fundamental attachment grounded in early socialisation" (Fenton, 1999, p. 102). It represents the idea that there is something inherently common to all members of a national or ethnic group. Alternately, a constructivist perspective that draws upon social constructivism (see Berger & Luckman, 1966) understands nation or ethnicity as concepts, brought by modern age and its specific conditions. This perspective refuses the premise of "something inherently common" to all members of ethnic or national groups. Instead, this perspective claims that these concepts are socially negotiated, and any link to "objective" (phenotypic) traits is illusory (Eriksen, 2019). Contemporary social sciences adopt the constructivist perspective and refuse the primordialist explanations of social behaviour. It means that ethnic boundaries do not necessarily correspond with cultural boundaries or with phenotypic characteristics (Eriksen, 2010, p. 41) but are constructed in everyday life within particular contexts.

The concept of ethnicity reveals a universal social phenomenon that may be applied all over the globe (Anthias, 1992; Eriksen, 2019). Depending on interpretations, it can be conceptualised as an informal political organisation, a product of colonialism, a functional boundary mechanism separating social groups, or an aspect of personal identity (Eriksen, 2019). In all these conceptualisations, ethnicity is a form of systematic categorisation of differences constituting the relational dichotomy of we versus the others. The systematic distinction between insiders and outsiders, i.e. in relation to we versus the others, is a "perennial feature of human groups" (Eriksen, 2019, p. 31). It is not possible to define and characterise ethnicity without a reference to relations with other social groups, which makes it a relational concept (Eriksen, 2010). Ethnic groups emerge and acquire their functional meaning through contact with other groups. These other groups are often seen as alien and exotic, and attention is paid to differences rather than to similarities. Stereotyping then generates expectations of typical behaviours and characteristics associated with particular ethnic groups (Eriksen, 2010). Stereotypes work on a principle of homogenisation and naturalisation, which means that typical behaviours and characteristics are expected to be representative of all members of a particular ethnic group.

Ethnicity is not only relational but also situational and contextual: ethnic characteristics have a different significance in different contexts and situations. Imagine a tourist of Chinese nationality and/or phenotypic traits. In Svalbard, such a tourist would probably be primarily seen as Asian. If the same person travels to Japan, their ethnicity will be primarily seen as Chinese. While in China, they will probably be primarily seen as Sichuan, for example. The significance of ethnic category is closely connected to the specific context and situation. Consequentially, in each of these contexts, the behaviour of the same person may be interpreted according to the particular expectations or stereotypes related to the type of ethnicity (e.g. typical Asian, Chinese, Sichuan, etc.). As Jonathan Okamura remarks, "a situational approach to ethnicity illuminates the fact that variability is the essence of ethnicity in its significance for the structuring of social relations in diverse situational contexts (Okamura, 1981, p. 463).

Given the fact that (Asian) ethnicity is relational, situational, and contextual, it is not a stable concept. Rather, it is a never-ending process of negotiations and interpretations of everyday situations, or as Thomas Hylland Eriksen puts it, "enduring and systematic communication of cultural differences between groups considering themselves to be distinct" (Eriksen, 2010, p. 69). From this perspective, ethnicity is not a tool to objectively describe a state of affairs, independent of individuals' interpretations and understandings. Rather than that, ethnicity is a flexible analytical tool that enables the capture and reflection of the significance and meaning in negotiating categorisations of social and cultural differences between people in a society. Or, as Floya Anthias refers specifically to the Asian ethnicity, "where the term Asian is used, it denotes a common label and experience. There are no 'accurate' definitions of a group" (Anthias, 1992, p. 423).

Research case and methological approach

In this chapter we interpret Asianness as a multi-dimensional ethnic concept that refers to "common labels and experiences" (Anthias 1992: 423) associated with tourists coming from Asia. In our understanding, Asianness as a concept is a combination of phenotypic traits, nationality, and geographic location, as well as being an externally ascribed category. This analytical approach enables us to show how the relational, situational, and contextual nature of the term Asian changes its significance with different contexts and situations pertaining to tour guides' planning processes in Svalbard. On the other hand, it limits its generalising potential, and our findings are not relevant for the Asian tourist as a group. This is in line with the constructivist perspective of situationality and contextuality of the ethnic concept, which rather than looking for accurate definitions aims to examine the mechanisms that assign common labels and meanings to ethnic categories, identities, and groups.

For this chapter, the research case is the process of planning guided tours. This process can be seen as a chain of decisions made by a tour guide who is supposed to deliver the desired result—a successful tour, through which everybody returns safely and satisfied. The guide's decisions are based on a partly conscious and partly unconscious series of assessments of tourist attributes. Such assessments serve to provide preliminary indicators of the character of a particular tourist group. This character further influences the organisation and tour design. The guide's assessments are necessarily based on their generalised interpretations of previous experiences with tourists and on particular interpretations of the characteristics and skills of tourists participating in the tour. This is the moment when common labelling, contextualisation, and situational aspects of ethnic relations take place. Therefore, we decided to focus on the planning process and to explore how interpretations drawn upon experiences, expectations, and common labels associated with Asianness influence tourist guides' planning.

Unlike most places in the world, guided tours are not an optional form of exploring the destination of Svalbard but the only legal option for unarmed tourists to experience the archipelago beyond the capital Longyearbyen. Such a situation is derived from Svalbard law, which requires any person leaving the inhabited settlements to be familiar with safety measures regarding polar bears and have an ability to use weaponry (Government of Norway, 2001). Each of the guided tours has a given activity (such as hiking, ATV, biking, kayaking, scooter driving, dog sledging, skiing, etc.), an area destination (i.e. particular mountain top; safari area; bird cliff, etc.), and a duration. From these offerings, tourists can choose and buy certain tours, which are delivered by guides. Since the tour operators do not limit the tourists' choice of the tour, they do state the level of difficulty of tours. Further, operators do not focus on specific tourist target groups; in practice, the group can vary in many aspects such as nationality, age, level of experience, etc. Part of a guide's duty is to work with the differences and dynamics within each group of tourists and to deliver the tour to the satisfaction of all. The guides thus learn and develop strategies and practices such as facilitating the slower members of the group, working with the mindset of individuals as well as with the atmosphere of the group, choosing appropriate paths according to the abilities of the least experienced, etc. The very last practice is to alter the goal of the tour while keeping the tourists satisfied.

The decision to use one of these practices is often made during the tour itself, but guides need to be prepared—to know the area and its possibilities; to know the weather and current terrain conditions; to have skills in communication with the individuals and the group; and to analyse tourists' abilities, skills, or experiences. Therefore, the guide has to prepare and gather available information about the natural conditions and character of the group, which might influence the design of the tour. If there is a significant difference of fitness within the group, the guide needs to try to overcome these differences, so none of the tourists are bored, cold, exhausted, or even injured. If there is a problem with equipment or clothing with one (or more) of the tourists, the guide needs to make sure the problem is fixed. There are many issues arising from the tourists, which a guide needs to work with, while ensuring safety, comfort, and the entertainment of the group. It is within this context that we describe in detail the process of tour planning by guides and throughout this process identify the significance of Asianness. We ask the following research questions: When and how does Asianness emerge during a guide's tour planning? What are the common labels and expectations of a typical Asian tourist among guides in Svalbard? What kind of role do the interpretations of guides play in the process of planning guided tours?

The data were obtained using the qualitative methods of participant observation and semi-structured interviews during Eva's ethnographic research into Svalbard tourism between 2016 and 2020. During this time, she participated in guided tours on a daily basis for two winter and three summer seasons with four different tour operators. She also conducted fifteen semi-structured interviews with guides. She has also worked in Svalbard as a guide herself, which provided her with many possibilities for sharing thoughts and seeing the guides in informal work settings during their tour planning. The data for this text cover a range of operators from small to larger scale and, most importantly, a variety of tours such as hiking, kayaking, boat trips, and dog mushing. All participant names have been changed, and other measures have been taken to ensure anonymity.

Asianness in the planning process of guides

As mentioned above, organising a tour in Svalbard requires specific measures. Due to the permanent danger of polar bears, safety on the tour is an absolute imperative. Guides have to employ all available information to anticipate and prevent potential problems. Therefore, we interpret the planning of the tour as a constant analytic process. This process starts when the first source of information is available, which is a list of bookings, and ends at the beginning of the tour itself when there is still time and possibility to turn back those participants who are deemed not able to complete the tour safely.

We identified and analytically differentiated three phases of a guide's planning process. It starts with a list of bookings, where initial assessments are made; continues with meeting the tourists and adjusting the initial assessment; and finally, observing the tourists' abilities on the tour itself. Analysing each of these phases separately, we focused on three aspects. The first aspect involved the common labels and expectations associated with Asian tourists among guides in Svalbard. The second aspect homed in on the role of interpretations during the process of planning guided tours. These interpretations drew upon experiences and expectations. The third aspect centred on the moments when Asianness emerged as a relevant tool for a guide's analysis.

Phase one: a list of bookings

Shortly before the tour, the guide gets a list of tourists who have booked the tour. On this list, there are names of the tourists, their booking number, date of birth, phone number, and pick up location. Based on this information, the guide assumes the character of the group and of the individuals within—and possible consequences for the tour. For example, an older tourist might be slower, or have troubles with terrain, while younger tourists might be rather fast and fit. If there are four people on the list who are rather old, and two who are considerably younger, there may be a difference in fitness amongst the tourists. Such a situation might lead to tensions on the tour, and the guide will have to think of ways or adjustments to satisfy all the tourists regardless of their differences.

> When you have a guest list, you always run a risk of them being terrible people; hopeless people; people that are not interested in what you are saying. The worst is when you have people that are really different from the rest of the group and it is difficult to make everyone happy. Then, you have to start compromising because you have to please different kinds of people.
>
> (Interview, Guide Esther)

One of the characteristics taken under consideration within the process of assessment is Asian ethnicity, which is derived from the "sound" of the name and from a phone number indicating the country of residence. If the phone number has a call prefix from India, Japan, or China, and/or if the name "sounds" like from one of the Asian languages or is considered exotic, an "Asian" ethnicity is ascribed to the person. The process is illustrated in the following quote:

> I look at the overall number of people and try to see if some of them know each other—then they would probably have the same surname or booking

> number. If there is a date of birth, I look at it as well. Then the name with the phone number, which suggest the nationality. It is basically just a glance over the list of bookings. On some trips, it is more important than others, but it gives you a suggestion of what to think about, what is the diversity of the group. After some time, you sort of learn, at least in general, how different names from different countries sound, and some of the most common country codes, or you google it to see the nationality.
>
> (Interview, Guide Joe)

Again, this scarce information about tourists is all the guide gets before the tour. In this context, a tourist's name and nationality are considered as a source of information, including certain expectations, that could be named as a process of stereotyping (Eriksen, 2010). In the following examples, we can see the kind of expectations that are typically associated with Asian tourists:

> Driving a car with the Norwegian tourists on the way to pick up the last two people, that are thought to be of Asian origin, Agnes, the guide expresses they expect troubles: "Maybe they will turn around soon and if not, I think the fitness difference between them and the 'North pole skiers' back in the car will be enormous. Sometimes it is hard work to make everybody happy, so no one is suffering from hard work and no one is being cold or bored or annoyed from walking too slow".
>
> (Fieldnotes, March 2017)

> I expect that the Scandinavian tourists have hiking experience and the Asian tourists I expect that they do not have experience in hiking or other outdoor activities.
>
> (Interview, Guide Agnes)

The guides' expectations of Asian tourists, based on the former's experiences, is that they are not very sporty; rather slow; not used to being in a rough terrain; not well equipped in terms of clothing; not very well self-informed about the destination; and generally inexperienced in outdoor activities. The assumption of the Asian origin of tourists in combination with high fitness levels of the already picked up tourists thus causes the expectation of high diversity in the group that will result in the impossibility to tailor the tour to all participants' needs and abilities, e.g. "to make everybody happy", as Agnes, the guide put it. On top of that, Asianness brings an expectation of poor English language skills.

The guide mostly considers those characteristics that possibly influence the tour such as skills in terrain, fitness, language skills, and equipment/clothing. Such a focus reflects two of the basic principles of guiding. The first one, we have already mentioned—it is to make the tourists happy and satisfied with the product for which they have paid. The other is rule number one—safety, which in Svalbard with polar bears and low temperatures is more significant than in many other destinations. Bringing back tourists safely is crucial. Therefore, the above-mentioned expected characteristics of Asian tourists are risk factors for every guide. Slow and ill-equipped Asian tourists thus mean not only more difficult work to give the tourists what they paid for but also create a potential threat to the safety of the group that must be taken seriously.

Guiding is a constant analytical process. The experience, expectations, and constant assessment of clients' abilities, limits, and desires are an inevitable part of it. Therefore, strong expectations associated with Asian tourists may have a significant influence on the planning process. A guide's attention is systematically focused on differences, which is one of the crucial mechanisms of ethnicity (Eriksen, 2010), because the differences within the group are what create potential difficulties or tensions whilst on tour. At this point, Asianness matters because the characteristics ascribed to it influence the work of the guide and might influence the character of the tour. Even before the tour starts, the guide may consider taking an easier trail or count on the option that the group will not be able to reach the mountain top, which is advertised as the tour's goal, just because of their assessment of tourist's abilities and fitness level, which are at this point influenced by ideas concerning Asian ethnicity. However, in this phase, these ideas are still more expectations and considerations rather than decisions.

Phase two: meeting the tourists

As much as the list provides certain indications about the character of the group and the individuals within it, the guide is aware that such indications are not invincible. Therefore, they further work with these expectations once they meet the tourist at the start of the tour. The guide talks to the tourists and based on their first impressions, previous expectations, and acquired information, they start to assess if there will be a need to adjust parts of the tour accordingly.

> When I hear that there's going to be Asian tourist, then I say ok, maybe we don't go to the summit but we spend more time lower, taking pictures, listening to a story, taking coffee, or whatever. You need to be much more customised for them, than just do what is written in the description of the trip. I always start to check the equipment. It might ruin the trip really easily, if they are dressed up poorly.
>
> (Interview, Guide Sam)

In this interview quote, we can see the moment at the start of the tour, when expectations are meeting with (first) impressions. Asian tourists are systematically categorised as "different", which is expressed in the guide Sam's words as "need to be much more customised". This requires the need to have alternative plans for the tour prepared and also "to check the equipment".

The moments of meeting the tourists have a significant influence on the guide and their tour planning. The guide sees more specifically in what condition the tourists are and what are their expectations but also assesses their level of experience based on clothing and general appearance or English skills. Meeting the tourists in person can confirm, disprove, or just modify the guide's expectations made on the basis of the list of bookings.

> I was picking up first at Marry Ann's, it was two Germans, older, but from Germany they have been hiking a lot and they had hiking shoes and all this gear that you need. They were prepared for the trip and they were kind of experienced. Then I went to the port, to get this Arab family and **of course** they were coming in sneakers. That day it was a little bit cold and the man in the family had a leather jacket and no gloves. **Of course**, no hats. So, I gave them gloves and hats and all the gear that I could give them.
>
> (Interview, Guide Halle; emphasis by the authors)

The guide's assessment of the tourists' appropriate clothing, language skills etc., is the process of negotiation and construction of ethnic relations. Asianness, as well as any other ethnic identity, is constructed in a relational, two-way process (Eriksen, 2010). One part of the construction is the common label of Asianness and the characteristics that a guide associates with it, which accumulates in their expectations as was expressed in the guide Halle's "of course". At the same time, these expectations have to be confirmed or declined by Asian tourists themselves, because if the expectations repeatedly do not meet with the reality, they will change. In situations like these, where the tourists fit into most of the expected Asian common labels (nationality, phenotypic features, lack of language skills or inadequate clothing), the characteristics considered typical for the Asian ethnicity are significant. This is how the homogenous and homogenising ideas and expectations about Asian tourists' characteristics are reproduced.

In phase two, expectations are confronted with first impressions. A guide's interpretations of Asian tourists become in this moment based not only on the expectations and experiences but also on the assessment of and concrete impression of the person with an Asian ethnicity. At this point, Asianness still matters, but it starts losing its relevance as the guide starts to confront their expectations with the people's general appearance, which gives them important input for their analysis of the tourist's abilities on the tour. However, the guide still can only assume what the actual performance of the tourists participating in the tour will be. The planning process is not over yet. It continues into phase three.

Phase Three: On the way—observing the abilities

In the very beginning, after all the tourists are gathered and briefed about the tour, a guide tries to get more detailed knowledge, for example, knee injuries or just current mood or general preference, and further works with that information during the tour. A guide also has to observe or "test" the tourists' abilities in the outdoors. Should some of the tourists have a lot of troubles at the very beginning of the tour, the guide will decide the pace, plan further breaks, choose a less difficult path or think of further decision points for the tour to complete within the time schedule.

> The first thing is always to see what clothing they have. However, you have poorly prepared guests who manage the trip and well-prepared guests that have to turn around. What I do is that I walk unnecessarily fast the first ten minutes of the trip to see who is able to follow and who isn't. This won't affect most people since they are well rested and ready for the day, so with a small break and tempo decrease afterwards they manage the rest of the trip fine. For the guests who generally would struggle with finishing the trip this is enough to make them realise that a full day of hiking isn't for them and they decide to turn around themselves.
>
> (Interview, Guide Esther)

Esther's reflection illustrates that in the last planning phase, interpretations made on the expectations and experiences meets the reality and are proved right or wrong. Even the poorly prepared tourists might be able to complete the tour with ease and no problems. Similarly, even Asian tourists may turn out to be good and experience hikers, as we can see in another field note:

> It was the fourth day of a multiday trip. After dinner, the guides were discussing the upcoming day—the plan was to hike up to the top of a steep mountain but also to prepare a less demanding alternative. Robin, the guide, thought out loud whether the numbers of tourists in each group would be approximately even: "People who I think would want to and be able to go to the top are Anja, Joe, Mike, Lene, Maria. I don't know about Sakura. She is so quiet. I don't even know if she understands or not. But she seems fit and capable. Quite surprising for such a tiny Japanese girl". The next day, the guides did offer the tourists to split and they agreed. Sakura, the rather small young lady, chose to go to the top of the mountain. After a demanding but smooth hike up the quite difficult mountain, all of the tourists made it all the way to the top, except Mike, who did not want to go the last most narrow 100 metres and he waited behind for us with the dog.
>
> (Fieldnotes, August 2018)

In the last phase, the initial expectations are definitively proven right or wrong, but the analytical work of a guide does not end. The guide continues to observe and to communicate with the tourists, and the input for the guide's analysis and decisions becomes more detailed compared to the first two phases of planning. The guide's interpretation becomes almost solely based on observation and communication with the individuals and the group. When they see a person is fit and well dressed, they are working with those specific characteristics, rather than with the expectations based on ethnic origin. As we could see in the case mentioned above, despite the expectations, the Japanese tourist Sakura made the difficult tour to the end, unlike the American tourist Mike, who in the beginning was not even considered as unable to finish the tour.

Ethnicity becomes irrelevant as a source of information for a guide's analytical work in the third phase. The guide's interpretations are based on direct observation of tourists' performances and previous expectations lose analytical relevance. As guides repeatedly state, they have to be prepared for the unexperienced tourists who are not fit and not able to assess their own abilities and especially their limits. But it does not mean that those expectations, associated with Asian tourists, are limiting the chance of Asian tourists if they actually prove to be able to complete a demanding tour. Unlike in the first two planning phases, Asianness is nearly absent in the last phase simply because it is not a meaningful source of information compared to the direct observation of a tourist's performance. As it was said, along with Jonathan Okamura (1981), ethnicity as a tool for creating interpretations is situational and flexible and changes its significance in diverse situational contexts accordingly.

Conclusion

In this chapter, we aimed to understand the term "Asianness" through the concept of ethnicity. Such a perspective provided us with a tool to focus on who in particular do people mean, when they refer to Asian tourists, and what expectations and common labels do they associate with this category. We examined the contextual and relational attributions of Asianness as an ethnic category. We showed how these attributions change their relevance during different phases of the tour planning process, specifically, how Asianness changes its significance as a source of information and an analytical tool for a tourist guide.

During the planning, the guide is constantly analysing and making their decisions according to their interpretations. Throughout this process, different sources of information are available. At the beginning, in phase one, a guide's expectations are drawn upon the "sound" of names and phone numbers indicating the country of residence in the list of bookings, and previous experiences with Asian tourists serve as a source of information. The interpretations of experiences suggest that common labels associated with Asian tourists are that they would be rather slow, not very sporty, and inexperienced in outdoor activities, often with poor English language skills. Therefore, they tend to be labelled as the type of tourists to whom one has to pay special attention and be prepared to adjust the tour according to their skills and abilities. The ethnic category Asianness is a strong clue in this phase.

In phase two, a new source of information appears. As the guide meets a tourist, they can see them and communicate with them. The expectations are confronted with first impressions. This creates new inputs for the expectations and interpretations of the guide which can be adjusted, and reformulated and most importantly can become more precise. As a result, Asianness becomes significantly less relevant as a source of information and analytical tool for the guide.

In phase three, Asianness is nearly absent. Compared to direct communication with tourists and observation of their performance during the tour, expectations associated with ethnicity become an irrelevant source of information. Generally, personal knowledge is a more relevant source of information and tends to replace ethnicity as a source of information.

We are not denying that in everyday life people still associate specific ethnic categories with certain common labels that exceed the praxis of tour guiding. Nevertheless, our aim was to show that guiding is an analytical process that systematically includes all kinds of available information and which they work with rather reflexively. What is often called stereotypes, e.g. homogenising expectations associated with the term "Asian" is a valuable source of information for guides in the initial phases of the planning process—simply because other information is not available. However, when guides meet tourists and especially when they are able to observe tourists' performances, they abandon their previous stereotyped expectations and ethnicity becomes irrelevant for their work and, consequently, the character of the guided tour. Therefore, we interpret the tourist guides' planning in Svalbard as a constant analytical process that is highly reflexive in its character.

Acknowledgements

This work was supported by the Ministry of Education of the Czech Republic as part of its scheme advancing research at universities under the project "Society in times of crisis and the institutional consequences of technology" number MUNI/A/1475/2020.

References

Anthias, F. (1992). Connecting race and ethnic phenomena. *Sociology*, *26*(3), 421–438.

Berger, P. L., & Luckmann, T. (1966). *The social construction of reality: A treatise in the sociology of knowledge*. London: Penguin Press.

Bhopal, R., Phillimore, P., & Kohli, H. (1991). Inappropriate use of the term 'Asian': An obstacle to ethnicity and health research. *Journal of Public Health*, *13*(44), 244–246.

Bott, E. (2018). Among the piranhas: The troubling lifespan of ethnic tropes in "tribal" tourism to Vietnam. *Journal of Sustainable Tourism*, *26*(8), 1291–1307.

Bruner, E., & Kirshenblatt-Gimblett, B. (1994). Maasai on the lawn: Tourist realism in East Africa. *Cultural Anthropology*, *9*(4), 435–470.

Cohen, E. (2016). Ethnic tourism in mainland Southeast Asia: The state of the art. *Tourism Recreation Research*, *41*(3), 232–245.

De la Maza, F. (2018). Tourism in Chile's indigenous territories: The impact of public policies and tourism value of indigenous culture. *Latin American and Caribbean Ethnic Studies*, *13*(1), 94–111.

Drew, E. M. (2011). Strategies for antiracist representation: Ethnic tourism guides in Chicago. *Journal of Tourism and Cultural Change*, *9*(2), 55–69.

Eriksen, T. H. (2010). *Ethnicity and nationalism: Anthropological perspectives. Third edition*. London: Pluto Press.

Eriksen, T. H. (2019). The epistemological status of the concept of ethnicity. *Anthropological Notebooks*, *25*(1), 27–36.

Fenton, S. (1999). *Ethnicity: Racism, class and culture*. London: Macmillan Press.

Government of Norway (2001). *Svalbard Environmental Protection Act, section 30a*. Retrieved from https://www.regjeringen.no/en/dokumenter/svalbard-environmental-protection-act/id173945/

Hall, M.C. (2009). 'A long and unfinished story?': Constructing and defining Asian regionalisms. In T. Winter, P. Teo, & T.C. Chang (Eds.), *Asia on tour: Exploring the rise of Asian tourism* (pp. 21–33). Abingdon: Routledge.

Lee, J., & Ramakrishnan, K. (2019). Who counts as Asian. *Ethnic and Racial Studies, 43*(10), 1733–1756.

Ocampo, A. (2016). *The Latinos of Asia: How Filipino Americans break the rules of race.* Stanford, CA: Stanford University Press.

Okamura, J. (1981). The situational ethnicity. *Ethnic and Racial Studies*, *4*(4), 452–465.

Özdemir, C., & Yolal, M. (2017). Cross-cultural tourist behavior: An examination of tourists' behavior in guided tours. *Tourism and Hospitality Research, 17*(3), 314–324.

Tonkin, E., McDonald, M., & Chapman, M. (1989). *History and Ethnicity*. London: Routledge.

Wood, R. E. (1998). Touristic ethnicity: A brief itinerary. *Ethnic and Racial Studies*, *21*(2), 218–241.

9 Thai mobility and consumerism at North Cape, Norway

Zooming in on a stone at a tourist site

Line Mathisen and Siri Ulfsdatter Søreng

Introduction

This chapter is about why a tourist site in northern Scandinavia has come to matter for Thais and how it continues to matter. Our story starts with a stone, *The Stone*, which is embedded in the rock surface at North Cape. The Stone bears the royal monogram of King Chulalongkorn (1853–1910), a Thai monarch who visited North Cape on 12 July 1907 and inscribed The Stone. We argue that The Stone has a particular place in stories about Thai world making and that it plays an essential role in today's Thai mobility and consumption of North Cape. Stories connected to The Stone inspire Thai mobility and enable the involvement of tourists to partake in consumption practices at the tourist site that facilitate the co-creation of value (Rakić & Chambers, 2012). North Cape as a tourist site draws on these practices by socially embedding tourists, who construct a variety of meanings and perceptions of the site during their visits (Ek & Hultman, 2008). Construction of meaning is anchored in tourist norms and values, which means that their basic values indirectly contribute to shaping mobility to and consumption of tourist sites (Hultman & Hall, 2012; Sørensen, Jensen, & Hagedorn-Rasmussen, 2018). This view of mobility combines an emic perspective of movement, which is closely connected to tourists' personal beliefs in daily life, with an etic perspective, which views tourism mobility as interlinked with other types of movements in time and space (Cohen & Cohen, 2015; Coles & Hall, 2006; Schwartz et al., 2012).

We seek to uncover the stories of The Stone by exploring the relations and meanings that empower it. We explore the stories to understand how The Stone connects to Thais' consumption and mobility of North Cape. This will, we argue, provide knowledge of The Stone as an actor in co-creation processes that have taken place across time and space and shed light on which values are co-created in these processes.

Tourist sites come to matter through individuals' engagement with practices and stories that by themselves, perhaps, do not create reasons to go, but when gaining power over time, create new practices and values—and thus new reasons to go. In this chapter, we describe how Thai tourists' reasons to go are co-created over time through multiple processes between individuals, artefacts, and nature. Inspired by Actor-Network Theory (Latour, 2005, 2012) and the

DOI: 10.4324/9781003039518-9

premise of value co-creation (Phi & Dredge, 2019; Vargo & Lusch, 2004), we aim to open The Stone's "black box" by tracing stories of practices and relations connected to it. In accordance with Latour (2005), who regards actors as humans and non-humans (artefacts) and links outcomes to how actors mobilise and enact to perform agency, we view The Stone as an actor embedded in a relational context with power to mobilise and enact agency to co-create value. Further, the notion of co-creation invites an approach focused on actors' interactions in culturally sensitive value co-creation processes (Akaka, Schau, & Vargo, 2013; Cannas, Argiolas, & Cabiddu, 2019; Edvardsson, Tronvoll, & Gruber, 2011). Combining these two perspectives enables us to open the black box of The Stone.

To answer our key question, *why a tourist site in northern Scandinavia has come to matter for Thais*, we start by introducing the tourist site where The Stone is located and unwrap the various co-creation processes connecting the stories of The Stone to a wider actor network. In addition to The Stone, another key-actor in our story about Thai mobility and consumerism to North Cape (NC) is King Chulalungkorn, who we will refer to as *The King*.

Contextualising our study

NC is a mountain plateau rising from the sea, about 300 meters high, located on the island Magerøya in the northernmost part of Norway. This is a dispersedly populated area (about 2,500 inhabitants). Most people live in the nearest city, Honningvåg, approximately 33 km away. NC is the land's end of mainland Europe. For centuries, it has been viewed as a natural highlight of a journey and an exotic site to experience the midnight sun (Favero, 2000; Herrero & Roseman, 2015; Tiffany, 1884). Reasons to go to NC have developed over centuries through stories of explorations and experiences from royals to the common traveller. For instance, in the 19th and parts of the 20th century, a journey to NC was an expedition by boat for adventure and prominent travellers searching for achievement, adventure, and inner harmony. Favero (2000) interprets such representations of NC as situated within a certain mythical framework that inspires contemporary (mass) tourism; tourists travel to search purity at the limits of civilisation to sustain their inner landscapes (Ives, Freeth, & Fischer, 2020). One prevailing assumption underpinning the production of myths is a tourist's search for authenticity (Chen & Chen, 2017; MacCannell, 2013). However, as tourists' inner landscapes, i.e. emotions, beliefs, and values are subject to the creation of meaning, tourist sites can only propose co-creation of value (Cannas et al., 2019; Prentice, 2001). To co-create, individuals participate actively (intellectually, physically, emotionally) in ongoing interactions to create an embodied experience of the self that can enhance the value of what they experience (Lugosi, 2014). This experience is linked to personal and cultural values (Kim, 2020; Schwartz & Bilsky, 1987; Schwartz et al., 2012), i.e. an individual's sense of self, their social and cultural heritage, and relations with other actors, artefacts and nature (Latour, 2012; Lugosi, 2014; Wang, 1999).

The first organised tours to NC date back to the 19th century. Over the last five decades, NC has become a site for mass tourism (Favero, 2000; Jacobsen, 2015). From the 1980s to the 1990s, visitor numbers increased; for example, in 1985, NC had about 100,000 visitors and 250,000 in 1994. Up until 2019, and notably prior to COVID-19, numbers of visitors remained stable and reflected similar numbers. Visitors were concentrated in the summer months, June to August, and arrived as individuals or in groups by cars and buses but mainly by cruise boats.

Most tourists visiting NC today are Europeans, of whom most are Germans. Asian visitors are only a small share of the total number of visitors (in 2019, about 6,000), but this number is slightly increasing. The relatively low number of Asian tourists reflects findings from Bui and Trupp (2019), who suggest that Asian mobility to Europe is relatively low, particularly amongst Thai tourists due to their financial situations. However, it is suggested that Thais who travel to Europe increasingly demonstrate consumption practices that connect to the personal value of status in their daily life (Bui & Trupp, 2019; Cohen & Cohen, 2015), i.e. status consumption (O'Cass & McEwen, 2004). Another important value shaping Thais' consumption and mobility to Europe is the cultural value of "sanuk", a desire to have a good time (Kanlayanasukho & Pearce, 2017). The most important destinations of Thai outbound travel in Europe are the United Kingdom, Germany, Netherlands, Denmark, Austria, and France. As these destinations refer to the first point of entry, they do not provide a thorough picture, for instance, of Thais visiting Norway (Bui & Trupp, 2014; Trupp & Bui, 2015).

There are no exact numbers for Thais visiting NC, as there is no registration of visitor nationalities at the site. However, numbers from Statistics Norway from 2019 show that Northern Norway experienced an increase in Thai visitors, particularly during late winter/early spring. It is difficult to make sense of these numbers, as they tell little of who is travelling, from where, and why. Yet, according to information from The Royal Thai Embassy, outbound Thai tourism to Norway is booming with about fifteen to twenty thousand Thais visiting Norway each year. Moreover, the number is increasing. Many of them also visit NC during their stay in Norway.

Method

To explore the role of The Stone for Thai mobility and consumption at North Cape, we draw on netnography, interviews with key informants, and observations. Netnography is an approach that enables the study of communication mediated through digital platforms. It is an underused but potentially fruitful method in tourism research, as it provides both an emic and etic point of view (Kozinets, 2015; Mkono & Markwell, 2014). We did not participate actively on these platforms; instead we explored posts linked to our research question. We collected data by using Google to search for a combination of the following keywords: "Initials Stone", "King Chulalongkorn", "North Cape", and "Thai tourism to Norway". In addition, we searched sites such as Facebook, TripAdvisor, and different blog communities focusing on these themes. Descriptive in-depth texts containing one or several of our key words were chosen. The web sites visited are established as public communication, meaning that consent is not viewed as necessary (Langer & Beckman, 2005). We also searched for literature linked to our themes concerning the tourist site and King Chulalongkorn, including The King's position today and research on Thai inbound/outbound travel. A limitation in our study is the fact that we are non-Thai speaking Norwegians.

Further, we interviewed The Royal Thai Embassy in Norway (hereafter called The Embassy) and the Director for NC Hall. The interviews were open-ended and provided information about Thai outbound tourism in Norway, the practice of Thai visitors at NC, and The King's meaning for Thais then and now. Finally, our observations of Thai practice at NC are linked to a visit to the tourist site in July 2019, therein shadowing Thai visitors experiencing the site. The data were analysed and themed according to the emerging stories connected to The Stone as they unfolded.

The King's consumption of North Cape

Thai mobilisation to and consumption of NC begun in 1907 with The King and his entourage visiting this site during his second journey in Europe. His first journey was in 1897. One may draw parallels between his European visits to the concept of "Grand Tours", which were practised by upper-class young Europeans, mostly men, accompanied by a chaperone (Trupp & Bui, 2015; Leiper, 2008). This kind of tourist behaviour could be understood as conspicuous consumption (Veblen, 1973) and status conception related to individuals' relations to owning possessions (O'Cass & McEwen, 2004). These trips often lasted for several months and were extremely expensive, like The King's European journey; The King had a reputation as a "big spender"' (Peleggi & Von Der Mehden, 2003). As a "Grand Tourist", The King visited multiple countries and cities where he met the ruling elite but also consumed art, culture, and leisure-oriented activities (Trupp & Bui, 2015). Based on letters that he sent home, his travel in 1907 became well known. These letters constitute a "travelogue", which was later published into a book, "Klai Ban" meaning far from home (Chulalongkorn, 1923, 1996). Embedded in Siamese cultural and ethical values and standards, formed by both Buddhist and Confucian norms (Bao, 1995), his letters tell much about his perception and reflection of both European and Siamese society.

The King reached NC by boat on 12 July 1907. In those days, NC was a well-known expedition goal restricted to prominent travellers (Tiffany, 1884) and probably a luxurious one for The King. In spite of the hardship of climbing up to the mountain plateau, The King describes NC as stunning: "At the top there was a flat land or a plateau although it was a bit wavy, and sometimes there were stones that were thick with ice (…). It was enchanting." (Chulalongkorn, 1996, p. 40 [our translation]). On the plateau, The King put The Stone in motion by signing it. He wrote:

> Phraya Cholayuth had brought along the usual tool used to carve my initials, but it was difficult to find a stone to carve them into. There was only one large

> piece of solid stone, all the rest were loose stones. We started by cleaning the surface. The carpenter and the carriers, five men in all, did the work. I wrote my initials and the year 1907 in Arabic numerals (…). When the work was over, we took a picture of it. I missed our horns. Usually we blow horns to mark when initials are engraved.
>
> (Chulalongkorn, 1996, p. 41)

The Stone is unique as it is the only stone outside of Thailand that The King ever signed (Figure 9.1).

The documentation of The King's visit (photos, letters, and the initial stone) are indeed influential artefacts empowering (future) value to be co-created when consuming the tourist site (Lugosi, 2014; Phi & Dredge, 2019). In the 1980s, The Stone was rediscovered by a Thai TV-team, travelling in the footstep of The King's European journey, and they turned it into a (symbolic) artefact of The King's presence at NC for Thais. The stories of The Stone started to circulate in Thailand, which triggered Thai authorities to initiate a Thai museum inside the NC Hall in respect and memory of The King. The Embassy played a crucial role. The King's great granddaughter, Princess Maha Chakri Sirindhorn, opened the museum in 1989. The photos at the Thai museum of The King's and the Princess' visits, together with statues of The King and Buddha, tell a story of continuity of the Thai (Royal) presence at the tourist site. Our findings show

Figure 9.1 King Chulalongkorn (to the right) at The Initial Stone on North Cape dated 12 July 1907. Source: National Archives of Thailand

that such continuity is also enshrined through official relations between Norway and Thailand, which go back to 1905, as Thailand was one of the first countries to recognise Norway's independence. Latour (2005) suggests that artefacts gain power through context, which means that contextual factors, e.g. stories, can trigger behavioural changes. The King's journey and The Stone gave birth to a story that since has prescribed particular behaviours across different contexts in time and space influencing Thais' present mobility and consumption of NC.

Thais' present consumption of The King

The stories of The King are central to understanding The Stone's influence on Thais' consumption of NC. Such consumption garners power due to The King's unique role in Thai society. The King was highly recognised and respected during his reign because he modernised Siamese society (Peleggi & Von Der Mehden, 2003). The Embassy explains that The King is considered one of Thailand's most beloved monarchs: "The reign of King Chulalongkorn was an age of extensive social and economic reform. Among the most important royal contributions was the announcement of the Slave Abolition Act, which was a clear indication of The King's great compassion for his people". In addition, The King managed to prevent Thailand from being colonised "by extending friendship and travelling extensively to western countries. Because of this, Thailand remains the only country in Southeast Asia that never to have been colonised". When The King passed at the age of 57, he was posthumously named "Somdetch Phra Piya Maharaj", which means "The Beloved Great King".

In Buddhist ideology, deceased kings can possess supernatural qualities, which make them "divine beings" or guardian angels" who actually may intervene in the world of humans as spirits (Stengs, 2009, p.19). The Beloved Great King is considered to hold the charismatic power of "Barami", which is tied to the "Ten Kingly Virtues" in Buddhism: charity, morality, self-sacrifice, rectitude, gentleness, self-restriction, non-anger, non-violence, forbearance, and non-obstruction (Stengs, 2003, p. 2). Connected to The King, through The Stone, we suggest that these values, amongst others, have meaning for Thais' consumption and mobility in NC today.

The King's popularity during his reign, and his "Barami", have led to the prevalence of a belief that those who link to him are empowered with a supernatural aura. During the late 1980s/early 1990s, a cult originating in the Bangkok middle class actually emerged around The King. The cult spread to include the working class and rural population (Peleggi & Von Der Mehden, 2003; Stengs, 2009). In Bangkok, the worshippers frequently visit *The Equestrian Statue*, believing that the Great King's spirit is present and that some of his Ten Kingly Virtues will rub off and give them good karma. The worshippers offer him luxury items along with candles and incense sticks, a common practice believed to bring good luck (Cohen & Cohen, 2015; Peleggi & Von Der Mehden, 2003). According to The Embassy, many Thais show their respects for the great monarch by placing wreaths at The Equestrian Statue on his Memorial Day (23 October). Similar donating practices also take place at The King's initials stones in Thailand. Our observations and interviews confirm similar practices at NC; Thai visitors donate artefacts at his statue inside the Thai museum and at The Stone. This suggests that Thais' consumption and mobility in their daily life interlink with patterns of movement outside their normal context (Cohen & Cohen, 2015; Latour, 2005). Moreover, the practice of showing The King respect is a practice empowered, shaped, and maintained through stories and artefacts across time and space. This is illustrated in NC by the first photographs, The Stone and The King's travel letters, and through Thai mass media and government agencies. For instance, in 2019, The Embassy ran a media campaign to promote the newly renovated Thai museum inside the NC Hall, for which The Embassy is responsible for upkeep. The Embassy organised a media team from Thailand to make a documentary of The King's visit to Norway and its importance for Thai-Norwegian relations. The media crew, together with the great-great-granddaughter of The King (Thanpuying Sirikitiya), re-traced the footsteps of The King in Norway. The documentary was launched on social media platforms (in Thai), "forging ancestral connections that link the past to the present", as described by The Embassy.

The Thais' present consumption of their Great Beloved King thus seems to rest on three value-based pillars of Thai society, which is the nation, the religion (Buddhism, but partly also Confucian norms [Bao, 1995]), and the monarchy (Johnsen, 2010; Murashima, 1988; Wiktorin, 2005, p.86). These three pillars serve to assert the popularity and sacredness of The King in Thai society. Furthermore, the deeply set values and beliefs connected to each pillar give power to the stories linked to The Stone and thus reinforce the attractiveness of NC for Thai visitors.

Thais consuming their own culture in North Cape?

Starting with The King's journey to Europe, Thai outbound travel remained restricted to a small elite of royal and other wealthy travellers (Cohen, 2014; Cohen & Cohen, 2015). Economic prosperity in the 1980s increased Thai mobility and consumption of tourist experiences; however, travel to Europe is still considered a luxury (Bui & Trupp, 2019; Trupp & Bui, 2015). Research on present Thai outbound travel in Europe seems somehow to confirm a continuity of the practice as described in The King's European journeys. According to The Embassy, many Thais still use The King's European travelogue ("Klai Ban") as a reference when travelling abroad, especially those who wish to re-trace his footsteps. Thai travel to Europe remains luxury travel, accessible to reserved middle- and high-income groups (Bui & Trupp, 2019; Trupp & Bui, 2015). This type of travel could be understood as a kind of conspicuous consumption (Veblen, 1973) and as a way of presenting "status" (O'Cass & McEwen, 2004). However, our findings suggest a more complex connection between values and meanings embedded in Thai society, which nurture Thais' consumption and movement.

Our findings indicate that The King's posthumous status as The Great Beloved King, which brings him a mythical presence and saint like qualities, inspire Thais to embed his values in their own lives. In Thai schools, The Embassy reports, children are taught about his European visits, as this period is considered as a major turning point for Siam in terms of politics, diplomatic relations, and modern technology development. In this manner, the state-driven connection of Thais' past to the present is institutionalised from an early age. In this chapter, we argue that such practices serve to create artefacts with the power to shape behaviours, such as the practice of showing The Great Beloved King respect. Furthermore, the cult that has emerged around The King shows a desire to (re)connect with a part of Thai culture deemed particularly virtuous. Herein lies the attraction of NC. According to The Embassy, NC has a historical significance for Thais, due to the uniqueness of The Initial Stone.

For the same reasons, consumption of NC might also be attractive for Thais living in Norway. Numbers from Statistics Norway (2020) show that 21,097 Norwegian inhabitants had a Thai background, mainly due to Thai-Norwegian intermarriage. Our findings show that local Thai associations had a role in the newly renovated Thai museum at NC, specifically, fundraising. Local Thai associations are often involved when The Embassy facilitates official visits for Thai delegations to Norway. This is in line with The Embassy's mandate to promote bilateral relations between Thailand and Norway and to assist Thai nationals living in Norway. During the past years, The Embassy has arranged for several visits for royal family members to NC. This emphasises the influence the Thai state has in facilitating Thais' mobility and consumption in NC, and their role in connecting Thai past practices to the present. Furthermore, monks from the Norwegian Thai Buddhist Association, with a Temple and monastic order in South Norway (Johnsen, 2010), are occasionally invited to hold ceremonies at NC in the Thai museum. This serves to indicate that the three value-based pillars of Thai society (the state, the monarchy, and Buddhism) are all embedded at the tourist site. In this manner, in NC, the Thai past is connected to the present through Thai incoming/international tourists and Thai local tourists, who are integrated in Norwegian local communities. Thus, Thais' consumption and mobility in NC can be traced back to The King's visit to NC, specifically, The Stone that was put in motion with his signature. The Stone became an actor in co-creating value to NC for Thais, making the tourist site a spot for consuming their own culture and history, therein showing their Great Beloved King respect. As Thai visitors at NC put it, "I am proud of King Chulalongkorn, who is involved in the history of the North Cape!", and "I am so proud to be Thai. The nation, the religion, and the king!" (posted on Facebook).

Closing comments

Prevailing perceptions of mobility and consumption of NC are heavily influenced by stories told by visitors and commercial and political actors. The stories can be considered as contexts for the creation and development of artefacts that suggest particular behaviours and as such, influence co-creation of value. In our study, the stories about The King's visit to NC are made visible or vivant through The Stone. The King's visit to Europe was essential to Siam and its peoples, and the stories about his visit remain important. The message in these stories is clearly connected to the value-based pillars of Thai society, and in this manner, we argue, the inner landscape of visitors' experiences of NC as a tourist site (Favero, 2000; Jacobsen, 2015; Tiffany, 1884). Further, these inner landscapes are characterised by values and beliefs that give meaning to and shape stories of the self. Our study suggests that for Thais, these stories are anchored in both personal and cultural values representing and connecting their past and present societies. Thus, when visiting the NC, viewing The Stone with The King's inscription, throwing in a penny, burning incense in the Thai museum, they benefit themselves and pay tribute to their culture and society in a broader sense. Their own practices become a part of a major story about the meaning of the Thai nation, religion, and the monarchy.

Further, from the perspective of value co-creation (Vargo & Lusch, 2004), our research highlights that tourists' mobility and consumption are linked to active involvement in emotional and physical activities. Our study highlights the connection between tourism consumption and mobility to the Thais' personal and cultural/religious values, i.e. their inner landscapes. Thus, our study shows that the value of NC as a tourist site is constructed through rather complex co-creation processes where stories (re)told over generations that resonate with individuals' inner landscapes play an essential part. Further, we believe that The Stone that bears The Great Beloved King's initials is the mediating force for Thai co-creation processes at NC. In this manner, the site holding The Stone represents Thai cultural heritage, as it embodies a shared bond with Thai society, connecting the past to the present, through culture, values, and traditions, and showing an index of personal and culturally embedded value(s). This raises questions linked to both what co-creation is, particularly in terms of what values are co-created at tourist sites and for whom. Further, our study illustrates a need to distinguish between personal and cultural values to fully understand the concept of value in terms of co-creation, consumption, and mobility in tourism. In the case of NC as a site for mass tourism, its significant value for Thais highlights issues linked to commercialisation, ownerships and governance. How should such values be commercialised? Alternatively, perhaps it is necessary to take a step back and ask whether they should be commercialised. Who should govern this heritage, and what consideration should be taken? We suggest that these questions should be addressed in future research in order to enable sustainable value creation in tourism for the future.

References

Akaka M. A., Schau H. J., & Vargo S. L. (2013). The co-creation of value-in-cultural-context. In *Consumer Culture Theory* (Vol. 15, pp. 265–284): Emerald Group Publishing Limited.

Bao, J. (1995). Sino-Tahi ethnic identity: Married daughters of China and daughters-in-law of Thailand. *Asian Journal of Social Science*, *23*(1), 57–77.

Bui, H. T., & Trupp, A. (2014). The development and diversity of Asian tourism in Europe: The case of Vienna. *International Journal of Tourism Sciences*, *14*(2), 1–17. doi:10.1080/15980634.2014.11434689

Bui, H. T., & Trupp, A. (2019). Asian tourism in Europe: Consumption, distinction, mobility, and diversity. *Tourism Recreation Research, 45*(1), 4–18. doi:10.1080/02508281.2019.1634305

Cannas, R., Argiolas, G., & Cabiddu, F. (2019). Fostering corporate sustainability in tourism management through social values within collective value co-creation processes. *Journal of Sustainable Tourism, 27*(1), 139–155. doi:10.1080/09669582.2018.1501053

Chen, J., & Chen, N. (2017). Beyond the everyday? Rethinking place meanings in tourism. *Tourism Geographies, 19*(1), 9–26. doi:10.1080/14616688.2016.1208677

Chulalongkorn. (1923) *Klai Ban*. Bangkok: Praephittya.

Chulalongkorn (1996). Reisebrev fra Norge 1907. Schibsted, Oslo. http://urn.nb.no/URN:NBN:no-nb_digibok_2008081100026

Cohen, E. (2014). The permutations of Thai Tourism. In K. Husa (Ed.), Southeast Asia Mobility Transitions – Recent Trends in Tourism and Migration. Abhandlungen zur Geographie und Regionalforschung 19, p. 237–261

Cohen, E., & Cohen, S. A. (2015). A mobilities approach to tourism from emerging world regions. *Current Issues in Tourism, 18*(1), 11–43. doi:10.1080/13683500.2014.898617

Coles, T., & Hall, M. (2006). The geography of tourism is dead. Long live geographies of tourism and mobility. *Current Issues in Tourism, 9*(4–5), 289–292.

Edvardsson, B., Tronvoll, B., & Gruber, T. (2011). Expanding understanding of service exchange and value co-creation: A social construction approach. *Journal of the Academy of Marketing Science, 39*(2), 327–339.

Ek, R., & Hultman, J. (2008). Sticky landscapes and smooth experiences: The biopower of tourism mobilities in the Öresund region. *Mobilities, 3*(2), 223–242.

Favero, P. (2000). " O'sole mio": Italian charter tourists' experiences of the Midnight Sun in North Cape, Norway. *Anthropological Quarterly*, 1–19.

Herrero, H. P., & Roseman, S. R. (2015). *The Tourism Imaginary and Pilgrimages to the Edges of the World* (Vol. 44): Channel View Publications.

Hultman, J., & Hall, C. M. (2012). Tourism place-making: Governance of locality in Sweden. *Annals of Tourism Research, 39*(2), 547–570.

Ives, C. D., Freeth, R. & Fischer, J. (2020). Inside-out sustainability: The neglect of inner worlds. *Ambio, 49*(1), 208–217.

Jacobsen, J. K. S. (2015). North Cape: In the land of the Midnight Sun. *The Tourism Imaginary and Pilgrimages to the Edges of the World, 44*, 120–140.

Johnsen, I. S. (2010). Wat Thai Norway: Buddhistisk praksis i et thailandsk tempel i Norge. Masteroppgave i religionshistorie, University of Oslo,https://www.duo.uio.no/handle/10852/23905

Kanlayanasukho, V., & Pearce P. L. (2017). Thai tourists on tour. The practice of designing successful tours. In P. L. Pearce & M.-Y. Wu (Eds.), *The World Meets Asian Tourists. Bridging Tourism Theory and Practice*. (Vol. 7). Bingley: Emerald.

Kim, M. (2020). A systematic literature review of the personal value orientation construct in hospitality and tourism literature. *International Journal of Hospitality Management, 89*, 102572. doi:https://doi.org/10.1016/j.ijhm.2020.102572

Kozinets, R. V. (2015). Netnography. *The International Encyclopedia of Digital Communication and Society*, 1–8.

Langer, R., & Beckman, S. C. (2005). Sensitive research topics: Netnography revisited. *Qualitative Market Research: An International Journal, 8*(2), 189–203. doi:10.1108/13522750510592454

Latour, B. (2005). *An Introduction to Actor-Network-Theory*. New York: Oxford University Press.

Latour, B. (2012). *We Have Never Been Modern*. Cambridge: Harvard University Press.

Leiper, N. (2008). A royal tour in 1897: An indicative example of tourism's impacts in the generating regions of whole tourism systems. *Tourism Culture & Communication*, *8*(3), 169–181. doi:10.3727/109830408786177461

Lugosi, P. (2014). Mobilising identity and culture in experience co-creation and venue operation. *Tourism Management*, *40*, 165–179. doi:https://doi.org/10.1016/j.tourman.2013.06.005

MacCannell, D. (2013). *The Tourist: A New Theory of the Leisure Class*: University of California Press.

Mkono, M., & Markwell, K. (2014). The application of netnography in tourism studies. *Annals of Tourism Research*, *48*, 289–291.

Murashima, E. (1988). The origin of modern official state ideology in Thailand. *Journal of Southeast Asian Studies*, *19*(1), 80–96. Retrieved from http://www.jstor.org/stable/20070993.

O'Cass Aron, & McEwen Emily. (2004). Exploring consumer status and conspicuous consumption. *J Consum Behav.*, *4*(1), 25–39.

Peleggi, M., c Von Der Mehden, F. R. (2003). Lord of things: The fashioning of the Siamese monarchy's modern image. *Pacific Affairs*, *76*(3), 499.

Phi, G., & Dredge, D. (2019). Critical issues in tourism co-creation. *Tourism Recreation Research*, *44*(3), 281–283.

Prentice, R. (2001). Experiential cultural tourism: Museums and the marketing of the new romanticism of evoked authenticity. *Museum Management and Curatorship*, *19*(1), 5–26. doi:10.1080/09647770100201901

Rakić, T., & Chambers, D. (2012). Rethinking the consumption of places. *Annals of Tourism Research*, *39*(3), 1612–1633.

Schwartz, S. H., & Bilsky, W. (1987). Toward a universal psychological structure of human values. *Journal of Personality and Social Psychology*, *53*(3), 550–562. doi:10.1037/0022-3514.53.3.550

Schwartz, S. H., Cieciuch, J., Vecchione, M., Davidov, E., Fischer, R., Beierlein, C., . . . Demirutku, K. (2012). Refining the theory of basic individual values. *Journal of personality and Social Psychology*, *103*(4), 663.

Stengs, I. (2003). *Worshipping the Great Moderniser: The Cult of King Chulalongkorn, Patron Saint of the Thai Middle Class*. Universiteit van Amsterdam [Host].

Stengs, I. (2009). *Worshipping the Great Moderniser: King Chulalongkorn, Patron Saint of the Thai Middle Class*: Nus Press.

Sørensen, F., Jensen, J. F., & Hagedorn-Rasmussen, P. (2018). Tourism place experience co-creation. In Kozak, M. & Kozak, N. (Eds.), *Tourist Behavior* (pp. 1–18). Cham: Springer.

Tiffany, C. C. (1884). Norway and the Midnight Sun. *Journal of the American Geographical Society of New York*, 273–310.

Trupp, A., & Bui, H. T. (2015). Thai outbound tourism to Austria: Trends and issues. *SDU Research Journal Humanities and Social Sciences*, *12*(1), 1–14. Retrieved from http://repository.usp.ac.fj/id/eprint/9486.

Vargo, S. L., & Lusch, R. F. (2004). Evolving to a new dominant logic for marketing. *Journal of marketing*, *68*(1), 1–17.

Veblen, T. (1973). *The Theory of the Leisure Class*. Boston: Houghton Mifflin.

Wang, N. (1999). Rethinking authenticity in tourism experience. *Annals of Tourism Research*, *26*(2), 349–370.

Wiktorin, P. (2005). *De villkorligt frigivna: relationen mellan munkar och lekfolk i ett nutida Thailand* (Vol. 15): Lund University.

10 Learning from an Arctic adventure guide

An exploratory study of emotional labour when leading Chinese consumers

Wilson Wai Yin Cheung

Introduction

According to the Adventure Travel Trade Association (ATTA), adventure tourism is one of the fastest-growing sectors of the tourism industry. Similar observations also can be found in Arctic destinations. Moreover, there has been a substantial increase in Asian tourists, largely represented by Chinese tourists, visiting the Arctic region in recent years (Cai et al., 2020; Jørgensen & Bertelsen, 2020; Lee, 2020; Xie & Tveterås, 2020). In 2018, China was the largest international tourists-generating country, with 149.72 million trips taken by outbound tourists, an increase of 14.7% from the preceding year (Travel China Guide, 2020). This rapid growth of Chinese visitors in the Arctic can be explained by the substantial growth of China's middle class with disposable income, as shown as Figure 10.1 (Cheung et al., 2019). While the Chinese outbound tourist market matures, characteristics of Chinese Arctic visitors are also shifting towards more personalised, exceptional experiences. Although the ATTA's (2013) adventure tourism market study report did not include outbound departures from Asia, Bannikin Travel and Tourism (BTT) investigated the profile of Chinese adventure travellers applying the same methodology as the ATTA. BTT found that 16.3% of Chinese travellers were adventure travellers and that the market was worth around USD 8.2 billion per year (Bannikin, 2019). As Asian participation in Arctic adventure tourism continues to grow in scale and scope, the interactions between frontline service workers and their Asian clients are increasingly important for reasons of both service quality as well as safety in wilderness settings. Such interactions place employees in emotionally complex situations, particularly in remote and isolated Arctic regions. However, apart from literature, which has often focused primarily on tourist experiences and behaviours, the few studies that have focused on guiding professionals have been reported by western scholars in English-based literature (Holyfield & Jonas, 2003; Sharpe, 2005). What is absent are studies related to adventure guiding of Asian travellers. That being said, a limited number of sociological studies have identified emotional and performative labour issues encountered by adventure guides in guide-to-client interactions (Holyfield & Jonas, 2003; Sharpe, 2005). Nevertheless, there is still a lack of knowledge about adventure guides related to leading culturally different clients and working

DOI: 10.4324/9781003039518-10

Figure 10.1 Chinese consumers display their national flag to mark their visit to this Arctic destination

in fully self-sufficient environments, such as polar regions. Accordingly, the purpose of this chapter is to provide an extended "insider's" perspective of Arctic adventure tourism guiding experiences. Based on the identified research gaps in extant literature, this book chapter focuses on the following: (1) how do Chinese visitors interpret adventure tourism in the Arctic? (2) how might the client-guide relationship affect such a process? (3) is there a specific consumption trait observable in Chinese visitors?

Unlike other tourist destinations, the Arctic region has a unique set of wilderness values, along with Nordic culture and local regulations, which create unique adventure experiences for visitors from different cultural backgrounds. Accordingly, an Arctic guide has a critical role as a gatekeeper for Arctic regions. Such a role is challenged not only by the physical risks in the region, such as polar bears, weather, and terrain, but also in fulfilling a demanding set of emotional expectations from employers, clients, and local authorities. The author of this chapter has been a polar expedition guide and leader, who has led more than 1,000 field days, especially for Chinese tourists. With this background, this chapter predominantly represents an "insider" perspective on tourists from a guide's experiences in either non-English speaking or Asian countries in the context of the Arctic adventure industry. Similar studies have been conducted by mountaineering guides, who present an "insider's" perspective on mountaineering adventure tourism (Beedie, 2003; Pomfret, 2006). Even though a number of studies have presented adventure guiding and identified emotional labour associated with guide-to-client interactions (Beedie, 2003; Holyfield & Jonas, 2003;

Pomfret, 2006; Sharpe, 2005); research on emotional labour in the specific context of cultural diversity and Arctic adventure tourist experiences are scarce.

Anecdotal accounts inform us that most of the Arctic destinations are not ready, or are unwilling, to cope with the constant growth of Chinese visitors as well as "their unique" consumption behaviours and culture. For instance, Chinese tourists are always associated with travelling in groups, engaging in limited interaction with local communities, and being shopping lovers. Such a lack of understanding or misinterpretation of the visitors not only leads to intensive stress and conflict between locals, visitors, and all other related stakeholders but also creates catastrophic crisis management related to environmental, cultural, and tourism resource aspects. In this regard, this chapter attempts to utilise my emotional labour on guiding Asian consumers' Arctic adventures as a foundation for further exploration of this topic. Although research on emotional labour has been active in some areas, the particular emotional demands of leisure-related jobs remain under-addressed (Ek et al., 2020). By understanding the emotional experience of Arctic adventure guiding associated with a constantly growing number of Asian consumers, outcomes will benefit adventure guides' well-being and contribute to a broader understanding of Asian consumers' expectations of Arctic adventure tourism.

Arctic adventure tourism and Asia

For the purposes of this chapter, the Arctic is defined as all of the regions north of 60° N like the islands in the Barents Sea such as Svalbard, Hopen, Jan Mayen and Franz Josef Land; Russia, specifically, the Murmansk Oblast, the Nenets; Alaska; Canada with northern Québec (Nunavik) and Labrador (Nunatsiavut); all of Greenland and Iceland; and the northernmost counties of Norway (Nordland, Troms, Finnmark), Sweden (Västerbotten and Norrbotten), and Finland (Lapland), whose boundaries lie closest to the Arctic Circle. Beyond its geographical scope, the general characteristics of Arctic destinations usually include remoteness, limited infrastructure as well as challenging logistics, fragile natural and cultural environmental encounters, strong seasonality owing to the extreme variations in daylight hours, and local weather necessitating flexible tourism operations.

Though the boundaries between adventurous activities and tourism are blurred, both explore the diffuse use of concepts such as wilderness, nature guiding, and adventurous activities (Rantala et al., 2018). While the United Nations World Tourism Organisation (UNWTO) has no definition of adventure tourism, the Adventure Travel Trade Association (ATTA) defines adventure tourism as an activity, which must involve at least two of the following three elements: (1) physical activity, (2) natural environment, and (3) cultural immersion. This mode of travelling is a way that presents opportunities for local communities to develop, while it takes into account visitors' understandings of what a responsible and sustainable adventure-based holiday involves. Huddart & Stott (2020) added two extra elements: self-exploration and connecting with self as benefits of adventure-based tourism. Further, they elaborated a more detailed definition of adventure tourism as having some of the following five elements: (1) some real or perceived risk related to physical activity, remoteness, or unfamiliar location; (2) occurring in a natural environment that is relatively unexplored and with minimal or no urban influences; (3) a physical activity that tests both mental and physical endurance; (4) cultural immersion; (5) a minimalism that supports eco-, responsible, and sustainable tourism and discourages mindless consumerism, although some of the luxury cruises that occur do not support this element (Huddart & Stott, 2020).

More so than most other destinations, Arctic adventure travellers are not only attracted to the remoteness and isolation but also to the authentic value and rare experiences offered in this northern polar region (Bislev & Smed, 2018; Martin & Ren, 2020). Accordingly, unlike traditional adventure tourism guiding, a unique set of sources of stress and emotions can be found in the Arctic guiding due to the unprecedented nature of this job. In order to manage responsible, environmentally friendly, and safe tourism in the Arctic, there are strict rules and regulations to which an Arctic guide must adhere in carrying out their work. This work includes consideration of vulnerable, natural environments, local cultures, and cultural remains, as well as challenging safety hazards at sea and on land. Arctic guides are experienced practitioners whose training and experiences are a product of the Polar Tourism Guides Association (PTGA), which validates their status and professions. Meanwhile, if the operation involves a cruise expedition ship, the guide or leader must follow the guidelines of the Association of Arctic Expedition Cruise Operators (AECO), which is an international association for expedition cruise operators operating in the Arctic. Before each season begins, each guide has to participate in an AECO field staff online assessment, which includes around 75–100 questions related to visiting sites, biosecurity, operational safety, and culture and wildlife safety, in order to refresh their knowledge of the region.

Apart from the Arctic guide's professional efforts, in order to conserve the Arctic's pristine environment and the safety of each activity, the traveller, as a participant, has to understand and follow the related guidelines and regulations. However, some studies have found that Chinese visitors have different kinds of understanding and interpretation of wilderness, adventure, and tourism (Cai et al., 2020; Cheung et al., 2019; Martin & Ren, 2020; Tin et al., 2016). In the last decade, the Arctic region saw a significant increase in Chinese travellers. However, sources of comparable visitor numbers data are limited due to data sets differing from one destination to another (Fay & Karlsdóttir, 2011; Huijbens & Lamers, 2017; Maher, 2017). Lee (2020) accessed a wide range of official statistical sources from European Arctic destinations to determine that there was an exponential growth in Asian tourists in general and Chinese tourists in particular. In the 21st century, formal Chinese Arctic tourism commenced around 2008 with one Chinese tour agent selling tickets for a European Arctic tour company ship (Huang et al., 2017). After that, there has been limited literature recording Asian travel patterns and their profile in the Arctic. Nevertheless, overall data from Arlt (2013) and Cheung et al. (2019) indicate the majority of Chinese people chose a group, mass-market package tour style to visit Arctic destinations due to lower prices, inferior quality, and language barriers. A detailed statistical record from the Icelandic Tourist Board (2020) showed that Chinese visitors in Iceland have increased from 5,500 persons (1.1% of the total number of visitors) in 2010 to 100,000 persons (5%) in 2019. China is one of the only three countries that emerged with more visitors to Iceland than in 2018, following Russia and Austria (Icelandic Tourist Board, 2020). For the same period of the time, due to social, economic, and environmental changes, the Chinese domestic outdoor tourism market started, which aligns with Icelandic statistics (Buckley & Buckley, 2016).

Despite traditional organised mass tourism being still common, the new wave of Chinese travellers seeks more open opportunities for individual experiences by venturing to remote areas with less human presence and higher environmental integrity (Arlt, 2013; Cheung et al., 2019). For instance, the vast majority of Chinese visitors travel on expedition cruise voyages to a number of Arctic "hot-spot" destinations like Svalbard, Greenland, Kamchatka, and the Canadian Arctic as well as the North Pole. The more intrepid ones also explore the Arctic interior by skiing or dog sledging to the North Pole or Greenland, climbing the highest peak—Denali—as well as participating in the Arctic marathon to the North Pole, Greenland. New forms of Arctic tourism activities drawing the attention of Chinese visitors include skydiving at the North Pole as well as travelling by a modified four-wheel drive vehicle to cross the Iceland ice cap and trans-Serbian terrains.

Although those activities fit into the term "adventure" in western culture, this is not the case in China. The closest term in Chinese is "戶外旅游", which can be directly translated into "outdoor tourism" (Buckley & Buckley, 2016). Yet, Chinese consumers attribute different associations to the elements of culture and understanding. In general, visiting a remote destination that is perceived as dangerous and exotic is represented by the general concept of "戶外旅游", which has no association with adventure activities or a detailed plan including risk assessment. Further, studies have reported that Chinese visitors have different understandings and interpretations of wilderness, adventure, and tourism (Cai et al., 2020; Cheung et al., 2019; Martin & Ren, 2020; Tin et al., 2016).

Autoethnography as a phenomenological research tool

There is a lack of previous research on the topic of "insider" insights into Arctic adventure tourism guiding for Chinese consumers. For that reason, since this study was exploratory in nature, an autoethnography methodology was employed in this chapter. This approach has been applied to a range of tourism experience studies, such as mountaineering (Beedie, 2003; Mackenzie & Kerr, 2012) and river guiding (Jonas et al., 2003; Mackenzie & Kerr, 2013). Unlike traditional ethnographic research, autoethnographic approaches require the researcher, who is already the "insider", to engage in autoethnographical enquiry and provide unique insights into personal and emotional lifeworlds (Noy 2007). The major difference between autoethnography and traditional methodologies is that researcher biases and reflexivity are openly acknowledged and discussed in autoethnography.

Consequently, an authentic, emotionally nuanced account of adventure tourism guiding experiences are conveyed and interpreted in a critical way. In autoethnographic research, two distinct approaches are recognised—evocative and analytic autoethnography (Anderson & Austin, 2012). An analytical autoethnographic approach employs traditional social science epistemologies and theoretical analyses, whereas an evocative autoethnographer presents the data by stimulating emotional empathy and perspective-taking using evocative communication techniques (Snow et al., 2003). The current chapter draws upon analytical autoethnographic techniques to combine evocative writing and a visible researcher-self with an explicit analytical agenda and places the researcher in the context with other group members, while conceptually analysing data through an established theoretical lens. It is hoped that this approach allows readers to empathise with the authenticity of emotions conveyed and thereby gain a more intuitive understanding of findings. The method used in this study satisfies Anderson's (2006) five key autoethnographic features, namely: complete member status of the researcher; analytic reflexivity; narrative visibility of the researcher's self; dialogue with informants beyond the self; and commitment to theoretical analysis.

Qualitative data analysis is dependent on the extent to which an investigator establishes sufficient credibility and trustworthiness. Therefore, a "thick description" of the primary researcher, data collection, and procedures are reported (Denzin, 1989, p.83). Data collection commenced in 2011, when I, the author, entered the role of a polar (both Arctic & Antarctica) expedition guide/leader. Since then, with my Chinese language skills and intensive polar industry experience, I have accumulated more than 1,000 polar field days. I have led over 70 expeditions for international scientists and visitors, especially Chinese-speaking consumers, in extremely remote areas, including the Antarctic continent and the Peninsula, Greenland, Kamchatka, Alaska, Norway, Iceland, Canada, the Arctic, and the Alps. I am the first Chinese person to obtain a PTGA "Senior Polar Guide Certificate". As a consequence, this chapter seeks to convey an authentic and emotionally nuanced account of my Arctic adventure tourism guiding experiences, while simultaneously viewing that entire experience through a critical lens.

Setting the scene, data collection, and analysis

During each Arctic season, I used to spend three to four months working with Chinese-speaking guests in the Circumpolar Northern regions. The group size varied from two-person private tours to over 200 cruise ship passengers with an average trip length of ten days. In some rare operations, such as climbing the highest peak—Denali—or skiing on the Iceland ice cap, the day length would be extended to 21 days. During this intensive working experience with Chinese clients, most of the data was collected in the form of my journal entries, emails, and field notes from conversations with fellow co-workers (mainly westerners), Chinese guests, and tour leaders since my entry into the Arctic guiding experience in 2011. Multiple data sources were used to triangulate and verify the consistency of my interpretations.

In order to enhance the credibility and validity of my observations, a structured vignette analysis was employed in this chapter, which has proved advantageous in revealing several layers of awareness in the process of cross-cultural analysis (Pitard, 2019). To achieve this research outcome, a six-step framework was developed: (1) context, (2) anecdote, (3) emotional response, (4) reflexivity, (5) strategies developed, (6) conclusive comments on layers (Pitard, 2019). With this framework, researchers allow themselves the liberty of writing freely while expressing their lived experiences within their anecdotes and only allowing themselves reflection and editing when writing analyses of the moments described in their vignettes (Pitard, 2015). This method empowers "individual vignettes to describe distinct experiences while connecting these stories through recounting the context and revealing new strategies developed to cater for the researcher's growing cultural awareness" (Pitard, 2015). For the purpose of pinpointing the outstanding cultural impact interactions regarding that which is culturally different, the researcher should place themselves back into the pre-reflective moment by using journal entries and photographs as prompts. In this case, the recall phase—anecdote—is the closest stage for a researcher to review the pre-reflective moment of a happening themselves. Afterwards, the process of reflexivity is applied to achieve an expansion of understanding in cultural interactions. Additionally, such vignettes, as in the case of my Arctic guiding experience, should stimulate an emotional response and provoke understanding from readers (Humphreys, 2005). With this emotional connection, the reader is conveyed to the moment of truthfulness of the researcher as well as the corresponding conclusion.

Further, a structured vignette analysis reinforces the necessity of all these research elements within autoethnographic writing; the following three anecdotes from my personal Arctic guiding experience attempt to illustrate the emotional labour (emotional response & reflexivity) in each scenario as well as a discussion related to each of the research questions (strategies developed and conclusive comments on layers).

Results and discussion

"A sip of adventure" by their own definition—how do Chinese visitors interpret adventure tourism in the Arctic?

Anecdote 1—sea kayaking in Greenland

Me: The entire local community is laughing at this kayaking group of 50 people from China. Regardless of how the locals react to them, the Chinese feel very proud of themselves and do not understand what the locals are laughing at. The Chinese group told everyone here that they are doing a sea kayaking expedition in a group. Indeed, I talked with some of the visitors and lead them to a souvenir shop before their one-hour sea kayaking "expedition" (Figure 10.2). They have no clue what they will do and where they are on the map. All they know is they are sea kayaking in a group with whales and icebergs for one hour with the powerboat shuttle next to them. Moreover, the group photographer will help them to

Figure 10.2 A group of sea kayakers visiting a souvenir shop before their one-hour sea kayaking "expedition" in Greenland by shuttle powerboat

capture a good picture for WeChat. They are very excited about this five-day sea-kayaking expedition. After they leave, the locals tell me that they do not understand what the Chinese are thinking or doing. This is because the Chinese tourists all speak in Chinese and travel in a group with their Chinese service provider without interacting with the locals. Sadly, for me, I understand Chinese, and the Chinese group continuously complain about how poor Greenlandic living is and how China should help them, even buy land to build a shopping mall. I am apprehensive about such an invasion, as it could vanquish Greenlandic culture entirely.

Emotional response and reflexivity (emotional labour)

Indeed, I felt very grateful to the locals for sharing their thoughts with me as I respect and talk to them. I fully comprehend the standpoint of the Greenlandic community. As a sea kayaking guide, we usually travel in a small group with a low profile so as to paddle along the coastline self-sufficiently. Honestly, at first, I could not imagine managing 50 people, who may be first-time kayakers and who only stay at a hotel every day, which is accessed by powerboat transfer, as a means to experience the uniqueness of Greenland. During my guiding interaction with them as their sea kayaking guide, I was confused about what I was doing during the activities, as everyone just wanted to take pictures on their kayak instead of paddling as planned. After one hour of their noise around, I sat on my kayak to listen to the silence of Greenland. It seems the Chinese find it hard to understand the connection of the landscape. They were just like kids visiting a theme park, having fun, taking pictures, and then leaving.

Regardless of the viewpoints of guides and locals, the Chinese visitors seemed very satisfied with their "expedition" experience based on their definition of activities perceived from within a Chinese cultural bubble—Asian food served and Chinese tour leader translation with limited cognition of the local culture and geography. In fact, in the small Greenlandic community with limited provision, hotel staff also find it challenging to accommodate Chinese visitor requests due to cultural differences and language barriers. Apart from unrespectful and zero tolerance to the Arctic community's culture, I strongly disagree with developing the Arctic community with financial investments in this last frontier of wilderness. Even though some western visitors are against Greenland's hunting and animal welfare culture, they still do not intend to transform the community by an investment approach. If such a mindset materialises in Greenland, I do not want to imagine how the future of the Arctic would look.

Strategies developed and conclusive comments on layers

Despite refusing to immerse in Greenland's culture, taking the western definition of adventure tourism from ATTA, the group of kayakers indeed pursued their own adventure by engaging in physical activity in the natural setting outside their normal environment to achieve their own well-being. However, one may ask whether the very notion of adventure performed by Chinese tourists is in accord with service providers (locals, guides) or not? As already mentioned, unlike western culture, the notion of "adventure" does not exist in China, and the closest associated idea is "outdoor tourism". In the last decade, due to the promise of exciting experiences and no hard-related technical skill requirements, outdoor tourism has gradually been commercialised and gained popularity among the Chinese general public. In general, this kind of tourism is mainly commercially-based, consisting of highly organised and industrialised guided services that offer high-volume low-difficulty products for unskilled clients.

Like the "softening" of adventure tourism that is also happening in Finnish Lapland, this raises new demands on local service providers due to the inadequate competencies and skills of the tourists (Rantala et al., 2018). Given that such adventures serve as commodities, which promise experiences, to succeed they need to be rationalised, standardised, efficient products, marketed sophisticatedly for profit (Smart, 1992). However, some Arctic destinations, like Greenland, which have limited infrastructure in wilderness settings, are difficult to "soften" for commercial consumption. The service provider cannot guarantee the presence of wildlife, suitable weather for activities, or secure low-risk conditions because of the nature of the wilderness. Moreover, limited perceptions of the cognitive and affective attributes of tourism activities are often found in general tourists from

Figure 10.3 Social media confines Chinese visitors' experience of the Arctic

China (Jin & Sparks, 2017). The very meaning for them of pursuing one activity does not relate to seeking enjoyment or experience. Instead, photography and a presence on social media is far more essential for tourism activities, as shown as Figure 10.3.

No matter which nationalities, these inherent incommensurable differences create unnecessary friction and struggles. Local definitions of adventure tourism may misalign with visitors' expectations resulting in conflict between what the visitor wants or does at a destination being viewed by locals as "inappropriate". Accordingly, service providers, especially in the Arctic region, have to consider how the term "adventure" resonates with Chinese consumers, even among other nationalities, and recognise that it has connotations that include differences in cultural interpretation.

The role of tour leader confines visitors' experiences and interpretations—how might the client-guide relationship affect such a process?

Anecdote 2—a Chinese charter aboard a Svalbard and Greenland expedition cruising ship

Me: the Chinese tour leader claims that he has intensive experience in the Arctic, but his to-do list for us shows that he knows nothing about local regulations and expedition style. He requests a circumnavigation of the whole of Svalbard in ten days plus a sea ice landing midseason to watch polar bears with drones flying without official permission, etc. For sure, we (expedition guides) reject such lists and do our job as best we can. Consequently, he is angry and complains to the captain that he has paid a lot of money and this is what his 200 guests have come for. Yet, while he talks to his VIPs in Chinese, he bad mouths our guiding ability and the lack of professionalism to cover the non-delivery of his list-of-things. As a result, all the visitors think we are very bad and do not trust our selection of landings or expedition plan. We are indeed super-exhausted, and my colleague complains to the visitors as well. However, there is no way to change the condition, as the Chinese tour leader has misled the majority of the visitors and subsequently has created a lot of trouble for us to manage. Meanwhile, most of the western crew members do not understand what the tour leader told them in Chinese.

Emotional response and reflexivity (emotional labour)

This is exhausting as well as a helpless situation for me as a middle person. As a polar expedition guide, we are passionate to share as well as preserve the best of Arctic landscapes and cultures for visitors. Our guiding experience and regional knowledge are key to ensure the quality of the visit experience for tourists. Personally, I believe that visiting the Arctic region can build the affective connection between visitors and the landscape. However, without trust from visitors and tour leaders, such a guiding outcome cannot be achieved. Moreover, our advice and safety instructions become difficult to enforce due to this relationship breakdown. In the end, our workload and stress reached extremely high levels. Additionally, Chinese tour leaders came up with their plan without considering the local knowledge and safety precautions, which made the expedition planning more challenging, as we have to follow their wills for travel.

As I am the middle person between both stakeholders because I can speak both English and Chinese, I fully understand both the difficulties and challenges for the industry. Bringing me into such a situation always adds extra stress and workload due to the dynamics of cultural misunderstanding and language barriers. The most stressful part is dealing with the majority of the guests, who have been misled by their tour leaders. I feel commiserations for them having paid a lot of money for their vacations without receiving what they wanted and ended up being unsatisfied when leaving the cruise. Yet, this outcome should not be reflected in disrespect of the staff and guides on board.

Strategies developed and conclusive comments on layers

Although this situation is slowly being changed by a new wave of Chinese tourists who travel in small groups independently, participating in mass tourism products is still common for the majority of the population in China because of the competitive price of exclusive Chinese cultural-based services (Arlt, 2013; Cheung et al., 2019). Meanwhile, last-minute booking (less than a month) is a key characteristic of Chinese tourists (Bannikin Travel & Tourism, 2019). However, unlike other common tourist destinations, visiting the Arctic requires adequate preparation beforehand for related activities. Thereby, mass package tourism and Chinese tour leaders still play a significant role in preparing their last-minute booked clients to reach the Arctic. Accordingly, Chinese tourists are more likely to rely on travel agents to provide information and describe the experiences they would obtain. In this respect, the quality of the travel agents, as well as the tour leader, are a vital part of defining the visitor's Arctic interpretation and experience.

Despite the tour leader orchestrating a group of people and showing them around a tourist site with rudimentary consideration of their welfare, the Arctic adventure guides lead tourists to engage more directly through adventure instead of visually consuming the Arctic as a destination. In the Arctic, safety is paramount, yet practical survival skills are key to the performance. Risk is associated with the hostile terrain in which the tourist relies on the guide for safety. In this instance, the intertwined relationship between the client and guide becomes more significant and cannot be replaced by a tour leader.

Taking a proactive approach and enhancing the cultural sensitivity of Arctic service providers and frontline staff is also key to avoiding conflict between visitor expectations and service approaches. Despite most of the Chinese travellers visiting the Arctic for its uniqueness, they still prefer to maintain the sameness of China in terms of home commodities, food, and language during their travels. Currently, a guided Arctic experience is always based on a western approach for the majority of consumers. However, the value and concept of "wilderness", "expedition", and "adventure" exemplify differences in interpretation and understandings based on Chinese culture (Cheung et al., 2019; Martin & Ren, 2020; Tin et al., 2016). Consequently, to ensure the safety of Arctic adventure tourist experiences as well as reduce the stress associated with guiding Asian consumers, and even other nationals, effective intercultural understanding and guiding techniques have to be implemented.

By consumption to token own social status—is there a specific consumption trait observable in Chinese visitors?

Anecdote 3—Kamchatka Expedition:

Me: this is the most uncomfortable, disgusting dining experience I have ever had with tourist guests. All of them show off how rich they are. One of them started to say to a crew member that he could buy the entire ship. Then, others share their insights/intentions to purchase the mines here and the port in Kamchatka to exploit the local resources further. Moreover, they always take more food than they need at the buffet. In the end, there are leftovers on the table, which get completely wasted.

Emotional response and reflexivity (emotional labour)

I felt very frustrated with my job duties. Am I introducing the beauty of the Arctic or contributing to the transformation of the Arctic as a destination? Obviously, this group of visitors does not feel interested in the natural history and local culture about which the Arctic guides are passionate. Unlike traditional Arctic destinations, such as Svalbard, Kamchatka, and the North-East passage, cruise ship tourism is a niche market in China. Such tourism often draws the attention of the *nouveau riche* to visit, who enjoy the resultant Chinese economic reformation. I am not surprised by their mindset of capitalism even during their vacation period. However, it seems they do not have the awareness of the special environmental values of Arctic regions. My passion to be an Arctic guide is to preserve the origin of this wilderness and reconnect the relationship between visitors and the landscape; however, my professional role cannot be presented in such a context. And, it is heartbreaking for me to witness how visitors spoil our last Arctic wildernesses.

Strategies developed and conclusive comments on layers

The uniqueness of the value of the Arctic is the main driver attracting most worldwide visitors to the region. Unlike ordinary visitors, the traveller from China has an intertwined relationship between travelling and investment. Wu et al. (2019) determined a positive relationship between tourism activities and house prices in first-tier cities of China (Beijing, Shanghai, and Guangzhou). Noteworthy was the perception of "wilderness" and that China's collectivist culture may offer an explanation regarding Chinese consumption behaviour during vacations. According to one cross-cultural study in the polar region by Tin et al. (2016), despite most national visitors' agreement that the wilderness is a place to be protected, a substantial proportion of Chinese respondents believed the wilderness is also a future resource for society. The newly emerging middle class of China, who come from rapid urbanisation and industrialisation areas, can afford the expenditure of visiting this niche tourism destination. Therefore, such cognition of landscape and indigenous culture may influence these visitors to consume the landscape as a resource during their Arctic visitation.

In a culture like China, social structures are highly stratified under a collectivist culture (Peterson & Hofstede, 2003). People care about their "place" in the social hierarchy and respect authority. Wuthnow et al. (1984) suggested that the more a society highlights economic status difference, the more it focuses upon symbolic goods that mark the differences. People who are more hierarchically oriented have strong needs to claim their social identity because one's position is mostly defined by one's economic advance. In this regard, earlier Veblen (1973) noticed that social status competition had an important influence on consumption. Conspicuous consumption is not only peculiar to rational economic people but also a manifestation of social status and face. With these points, overconsumption of food carries the meaning of more than necessary, but also for Chinese "face culture" of social status, which results in excess food waste in the Arctic.

Conclusions

With more Chinese tourists flocking into the Arctic region and more development of tourism infrastructure and facilities, tourism activities may negatively affect the region's wilderness character and the cultural integrity of local communities. In this chapter, drawing on the perspectives of a guide and service provider point of view, struggles associated with catering to Chinese visitors in the Arctic were exemplified. While the three vignettes that were analysed overviewed national cultural differences and their stereotypical consumption behaviours, a major dilemma does exist in the current approach to Arctic tourism. Yet, China is currently undergoing drastic changes in social structure and consumer spending patterns due to a strong economic drive. Consequently, the trend of a new wave of Chinese tourists may demand alternative supply from Arctic destinations. Conversely, the new wave likes to travel in small groups and seek personal, unique experiences within local landscapes and culture connections.

Under the current state of capitalism, the emotion-inducing behaviour of service staff members is part of the service commodity. And, in the Arctic, there is no exception. Guiding in the last frontier of the northern wilderness has left behind polar culture. The duty of an Arctic adventure guide is not only to ensure the safety of clients as well as wildlife but also to preserve Arctic values. Guides have critical roles as gatekeepers who follow local regulations and legal frameworks in the execution of tourism activities. Apart from those responsibilities, given the changing dynamic of Arctic visitors from the Far East, guides may be operating within unfamiliar cultural contexts. Such contexts generally contribute to more stress and result in increased psychological adjustment problems. However, by enhancing awareness and understanding of diversity in cultural perspectives of the Arctic and focusing on balanced management as an outcome, the uniqueness value of the Arctic region may be retained.

Acknowledgements

A very special thanks to Professor Young-Sook Lee for her kindness and professional guidance throughout this entire book project. I would like to recognise and thank Professor Patrick T. Maher for connecting me with this valuable project. Finally, I am also grateful in receiving encouragement and help from Dr Thomas Bauer. I have been very much inspired by his research and sharing of polar studies. Besides, I would also like to thank Mr Nil Rodes i Llorens and Miss Maja Zimmermann, who have supported me during the COVID-19 pandemic. Additionally, my special thanks to Professor Gayle Jennings for her patient and invaluable assistance in proofreading this chapter.

References

Adventure Travel Trade Association. (2013). *Adventure Tourism Market Study 2013*. Retrieved from http://www.fairtrade.travel/source/websites/fairtrade/documents/ATTA_-_Adventure_Tourism_Market_Study_2013.pdf

Anderson, L. (2006). Analytic autoethnography. *Journal of Contemporary Ethnography, 35*(4), 373–395. doi:10.1177/0891241605280449

Anderson, L., & Austin, M. (2012). Auto-ethnography in leisure studies. *Leisure Studies, 31*(2), 131–146. doi:10.1080/02614367.2011.599069

Arlt, W. G. (2013). The second wave of Chinese outbound tourism. *Tourism Planning & Development, 10*(2), 126–133. doi:10.1080/21568316.2013.800350

Bannikin Travel & Tourism. (2019). *China Adventure Tourism Market Study*. Retrieved from https://bannikin.com/wp-content/uploads/2019/06/Bannikin-China-Adventure-Tourism-Report.pdf

Beedie, P. (2003). Mountain guiding and adventure tourism: Reflections on the choreography of the experience. *Leisure Studies, 22*(2), 147–167.

Bislev, A., & Smed, K. (2018). When will the iceberg melt? Narrating the Arctic among Chinese and Danish tourists aboard a cruise ship in Greenland. *China & the Arctic*, 2018, 29–42.

Buckley, R., & Buckley, R. (2016). 中国户外旅游回顾与展望——基于外国学者视角①. 地理科学进展, *35*(6), 665–678.

Cai, Y., Ma, J., & Lee, Y.-S. (2020). How do Chinese travelers experience the Arctic? Insights from a hedonic and eudaimonic perspective. *Scandinavian Journal of Hospitality and Tourism, 20*(2), 144–165.

Cheung, W. (wai Y., Bauer, T., & Deng, J. (2019). The growth of Chinese tourism to Antarctica: A profile of their connectedness to nature, motivations, and perceptions. *The Polar Journal, 9*(1), 197–213.

Denzin, N. K. (1989). *Interpretive Interactionism*. Newbury Park, CA: Sage.

Ek, R., Hardy, A., Larson, M., & Ooi, C. S. (2020). The emotional labor of the co-created tourism experience. In S. K. Dixit, (Ed.), *The Routledge Handbook of Tourism Experience Management and Marketing* (pp. 550–559). London, UK: Routledge.

Fay, G., & Karlsdóttir, A. (2011). Social indicators for Arctic tourism: Observing trends and assessing data. *Polar Geography, 34*(1–2), 63–86.

Holyfield, L., & Jonas, L. (2003). From river god to research grunt: Identity, emotions, and the river guide. *Symbolic Interaction, 26*(2), 285–306. doi:10.1525/si.2003.26.2.285

Huang, M., Tang, C., & Weaver, D. B. (2017). The Arctic tourism experience from an evolving Chinese perspective. In Y. S. Lee, D. B. Weaver, & N. K. Prebensen (Eds.), *Arctic Tourism Experiences: Production, Consumption and Sustainability*, (pp. 89–99). Wallingford, UK: CABI.

Huddart, D., & Stott, T. (2020). What Is adventure tourism?. In D. Huddart, & T. Stott (Eds.), edn] *Adventure Tourism*, (pp. 1–9). Cham, Switzerland: Springer International Publishing.

Huijbens, E., & Lamers, M. (2017). Sustainable tourism and natural resource conservation in the Polar regions: An Editorial. *Resources, 6*(3), 45.

Humphreys, M. (2005). Getting personal: Reflexivity and autoethnographic vignettes. In *Qualitative Inquiry, 11*(6), 840–860. doi:10.1177/1077800404269425

Icelandic Tourist Board. (2020). *Numbers of foreign visitors*. https://www.ferdamalastofa.is/en/recearch-and-statistics/numbers-of-foreign-visitors

Jin, X, & Sparks, B. (2017). Barriers to offering special interest tour products to the Chinese outbound group market. *Tourism Management, 59*, 205–215. doi:10.1016/j.tourman.2016.08.007

Jonas, L. M., Stewart, W. P., & Larkin, K. W. (2003). Encountering Heidi: Audiences for a wilderness adventurer identity. *Journal of Contemporary Ethnography, 32*(4), 403–431.

Jørgensen, M. T., & Bertelsen, R. G. (2020). Chinese tourism in the Nordic Arctic – opportunities beyond the economic. *Scandinavian Journal of Hospitality and Tourism, 20*(2), 166–177.

Lee, Y.-S. (2020). Asia and Arctic tourism. *Scandinavian Journal of Hospitality and Tourism, 20*(2), 105–109.

Mackenzie, S. H., & Kerr, J. H. (2012). A (mis)guided adventure tourism experience: An autoethnographic analysis of mountaineering in Bolivia. *Journal of Sport & Tourism, 17*(2), 125–144). doi:10.1080/14775085.2012.729901

Mackenzie, S. H., & Kerr, J. H. (2013). Stress and emotions at work: An adventure tourism guide's experiences. *Tourism Management, 36*, 3–14. doi:10.1016/j.tourman.2012.10.018

Maher, P. T. (2017). Tourism futures in the Arctic. In K. Latola & H. Savela (eds.), *The Interconnected Arctic — UArctic Congress 2016* (Vol. 4, pp. 213–220). Cham, Switzerland: Springer International Publishing.

Martin, N., & Ren, C. (2020). Adventurous Arctic encounters? Exploring Chinese adventure tourism. *Scandinavian Journal of Hospitality and Tourism, 20*(2), 126–143.

Noy, C. (2007). The language (s) of the tourist experience: An autoethnography of the poetic tourist. In I. Ateljevic, A. Pritchard, & N. Morgan (Eds.), *The Critical Turn in Tourism Studies: Innovative Research Methodologies* (pp. 349–370). Amsterdam: Elsevier.

Peterson, M. F., & Hofstede, G. (2003). Culture's consequences: Comparing values, behaviors, institutions, and organizations across nations. *Administrative Science Quarterly, 48*(1), 127. doi:10.2307/3556622

Pitard, J. (2015). Using vignettes within autoethnography to explore layers of cross-cultural awareness as a teacher. *Forum Qualitative Sozialforschung/Forum: Qualitative Social Research, 17*(1), Art. 11. doi: 10.17169/fqs-17.1.2393

Pitard, J. (2019). Autoethnography as a phenomenological tool: Connecting the personal to the cultural. In: P. Liamputtong (Ed)., *Handbook of Research Methods in Health Social Sciences* (pp. 1829–1845). Singapore: Springer. doi:10.1007/978-981-10-2779-6_48-1

Pomfret, G. (2006). Mountaineering adventure tourists: A conceptual framework for research. *Tourism Management, 27*(1), 113–123). doi:10.1016/j.tourman.2004.08.003

Rantala, O., Hallikainen, V., Ilola, H., & Tuulentie, S. (2018). The softening of adventure tourism. *Scandinavian Journal of Hospitality and Tourism, 18*(4), 343–361.

Sharpe, E. K. (2005). "Going above and beyond:" The emotional labor of adventure guides. *Journal of Leisure Research, 37*(1) 29–50. doi:10.1080/00222216.2005.11950039

Smart, B. (1992). *Modern Conditions, Postmodern Controversies*. London, UK: Routledge.

Snow, D. A., Morrill, C., & Anderson, L. (2003). Elaborating analytic ethnography. *Ethnography, 4*(2), 181–200. doi:10.1177/14661381030042002

Tin, T., Summerson, R., & Yang, H. R. (2016). Wilderness or pure land: Tourists' perceptions of Antarctica. *The Polar Journal, 6*(2), 307–327.

Travel China Guide. (1998–2020). *2018 China tourism facts & figures*. Travel China Guide. Retrieved from https://www.travelchinaguide.com/tourism/2018statistics/

Veblen, T. (1973). *The Theory of the Leisure Class*. Boston, The USA: Houghton Mifflin.

Wu, T.-P., Wu, H.-C., Chu, C.-C., & Deng, R. (2019). Tourism and house prices: A wavelet analysis. *Journal of China Tourism Research, 68*, 1–16.

Wuthnow, R., Hunter, J. D., Bergesen, A., & Kurzweil, E. (1984). The cultural anthropology of Mary Douglas. *Cultural Analysis*, 77–132.

Xie, J., & Tveterås, S. (2020). Economic drivers for the Chinese tourists. *Scandinavian Journal of Hospitality and Tourism, 20*(2), 110–125.

11 The Chinese Riddle

Tourism, China, and Svalbard

Zdenka Sokolíčková

Opening scene

It is 6 March 2020; I am sitting at Huset, a traditional venue for cultural events in Longyearbyen, Svalbard. The hall is full and spirits are high. A group of local artists is performing a satiric revue that summarises the year that has passed. One of the sketches is about Chinese tourists. The singers are dressed in large insulated jackets and are wearing face masks, moving like confused robots. At the end of the sketch, one of the singers drags a piece of luggage across the stage area while another asks where she is headed. “To China”, she replies. “The Chinese are ok, I am sure, but just as long as they stay home where they belong”. I laugh along with the others and think: “I have an opening for my chapter”.

Understanding Chinese tourists on Svalbard: empirically focused and anthropologically approached

Broad in focus and narrow in locale, this chapter explores how tourism, China, and Svalbard are interwoven. It zooms out and looks at the geopolitical, environmental, economic, and social context, but it also zooms in and studies local perceptions and lived experiences. In the stories encountered in the field, problematic areas emerge and lead to findings related to marketing and communication, racism, and (dis)ability to adapt to a rapidly changing and geopolitically tense world. The strength of social anthropology lies within the narration of how “small places” deal with “large issues” (Eriksen, 2015). My aim is to capture the phenomenon of Chinese tourism as a lively and constantly evolving process, something that embodies the paradoxical tension when Svalbard is placed on the global-local nexus, including identification of challenging issues and possible ways out.

Informing this chapter is an empirical study of a recent phenomenon—Chinese visitors in Svalbard. I have spent 20 months in Longyearbyen as a social anthropologist and my methods include in-depth interviews, focus groups, participant and nonparticipant observation, and auto-ethnography. Research participants that I met between February and September 2020 for the purpose of this study, be it physically or digitally, were people who worked with or even targeted exclusively

DOI: 10.4324/9781003039518-11

Chinese clients and employees of relevant stakeholders such as Visit Svalbard, Visit Norway, Association of Arctic Expedition Cruise Operators, AECO, or Hurtigruten Svalbard. The dataset also includes informal conversations with guides, people working in the service industry, a sinologist on an artistic residence, a Chinese influencer on a business trip, and last but not least *in situ* observations, statistics provided by Visit Svalbard and online sources. During the final phase of writing, preliminary results were made available to the wide public in Longyearbyen through a presentation and discussion in the library (in English), a chronicle in the local newspaper (Sokolickova, 2020), and a dissemination event targeting local tourism markets organised by Visit Svalbard (both in Norwegian). Making ongoing and unpublished research participative and available to those impacted is an important ethical prerequisite here.

Svalbard in a nutshell

Svalbard is both extraordinary and symptomatic when it comes to current trends in the changing Arctic. The archipelago of more than 61,000 km^2 is situated between latitudes 74° to 81° north, and from longitudes 10° to 35° east in the Barents Sea. It is often depicted as uninhabitable and without any Indigenous population. The territory was discovered in the late 16th century by a Dutch explorer, Willem Barentsz. During the 17th, 18th, and 19th centuries, the archipelago was sparsely inhabited by people involved in all sorts of extractive industries of early modern history. These people included Dutch, British, French, and Danish sealers and whalers, Russian trappers, and eventually also Norwegian walrus hunters, who came and left again. Whaling faded out in the mid-19th century, and a few decades later, large coal deposits were discovered. Norwegian, British, and American interests in the territory's potential started clashing. Norway succeeded diplomatically in drafting and signing the Spitsbergen Treaty in 1920, which came into effect in 1925. Among many other countries, the Treaty was also signed by the Republic of China. The document, nowadays known as the Svalbard Treaty has no analogies whatsoever. Its goal is to ensure peace and environmental protection through Norwegian governance, granting citizens of countries that ratified the Treaty the right to access Svalbard. It bans activities with war-like purposes and protects the rights of signatories in terms of fishing, hunting, and gaining mineral resources. Neither tourism nor science is explicitly mentioned in the Treaty. Norway and Russia managed to establish and preserve the costly coal mining industry in the geopolitically strategic territory. In the early 20th century, several settlements were founded; Longyearbyen was one of them (Arlov, 2003).

In Svalbard, the speed of change since the end of the Cold War seems to have accelerated. Not only has this part of the Arctic started warming up much faster than the rest of the planet, with an average increase of 5°C since 1970s, it also became very much connected to the outer world, quickly becoming "too full and too uneven" (Eriksen, 2016). By the late 1980s, tourism became a promising industry. The airport established in 1975 contributed to making the island accessible all year round. Longyearbyen is no longer a Norwegian village of a few hundred inhabitants, who are mostly miners, their families, industry managers, and a handful of state employees. The town is economically dependent on the Norwegian state budget, as it has always been throughout its short history, but it is also dependent on income from tourism. The tourist industry has contributed to the substantial growth of population and urban infrastructure. It has created numerous jobs and has been attracting both visitors and residents from all around the world. In September 2020, there were about 2,400 inhabitants drawn from 54 countries, with the three biggest minorities coming from Thailand, Sweden, and the Philippines.

The double bind of tourism in Svalbard

There is a fundamental dilemma embedded in the issue of tourism in Svalbard. Tourism is by no means a new activity in the archipelago; hyperbolically speaking, early polar expeditions can be deemed as a specific form of tourism. Terms such as mass tourism or over-tourism have been used lately (Saville, 2019a; Dvorak, 2020), and contradictions related to the presumed sustainability of tourism businesses have been widely studied (Van Bets et al., 2017). The local debate about the right balance to strike in tourism is lively, as it has multiple effects, and not all of them are positive. From an environmental perspective, it can both contribute to a better knowledge and a tighter bond of visitors to the unique ecosystem. Yet, Svalbard cannot do without frequent and cheap flights and a certain volume of visitors, which again creates further pressure on the environment (both in the settlements and out in the terrain, be it on land or sea). Economically, tourism contributes to the budget of the Community Council (e.g. through the income of Longyearbyen Port), and it creates local funding opportunities (e.g. in the form of Svalbard Environmental Protection Fund receiving 150 NOK from each flight ticket purchased to Svalbard or *korkpenger* from selling alcohol, which in turn is invested into local NGOs). Above all, it is the main source of income for hundreds of Longyearbyen residents. Forty per cent of *årsverk*, a workload equivalent to about 600 full-time all-year-round jobs available in Longyearbyen, is related to tourism. It also means the need to invest more in infrastructure that "makes the meeting between a tourist and a local less painful", as it was put at the Optimal Tourism Balance Workshop held in September 2019. There is no doubt about tourism making Svalbard, or in every case Longyearbyen, more global and vibrant. Without all the jobs that tourism generates, many of which are only seasonal, part-time, or suitable for freelancers, a large part of the international segment of Longyearbyen's population would never manage to afford the exorbitant costs of living on the fascinating island of Spitsbergen. The accelerating change that tourism triggered in Svalbard is multi-layered, together with other global phenomena such as climate change, migration, and geopolitical tensions. "Cross-scale" change seems to be "the constant" for Svalbard in the nearest future (Hovelsrud, Kaltenborn, & Olsen, 2020; Kaltenborn, Østreng, & Hovelsrud, 2020). When researching about tourism (Viken, 2006; Kelman, Rauken, & Hovelsrud, 2012; Hindley & Font, 2018; Saville, 2019b) in places where human beings cannot survive without the support of environmentally (so far) unfriendly technology, one inevitably ends up asking the question: why entertain tourists at all? Sustainability is a concept that needs to take environmental, economic, and social impacts into consideration. In Svalbard, the dilemma is even more profound, and many of my participants tabled the rhetorical question: why do people live in Svalbard?

"When will China come to Svalbard?" The unknown, vast, and changing Chinese market

In 2014, there were 497 Chinese guest nights registered in Svalbard, which would mean only about 200 Chinese citizens visiting the archipelago that year. Northern Norway and other Arctic destinations have lately been receiving numerous Asian and especially Chinese tourists (Wu & Chang, 2020; Xie & Tveterås, 2020). Such a shift did not become a topic of discussion in Svalbard until numbers went up from 3,142 guest nights in 2018 to 5,842 in 2019. The most striking increase occurred during the months of July and August. With the average length of stay equal to 2.4 days, 2019 numbers indicate that the stream from China to Svalbard has grown ten times within five years, while the total volume of guest nights has less than doubled during the same period. The numbers are still relatively low, but the speed of increase was high in the pre-corona world. "I remember that somebody at the airport, a stewardess, told me in 2018 that during one period, they called the planes coming from Oslo to Longyearbyen the 'Beijing Express'", a Visit Svalbard employee recalled. Educated guesses by personnel at Longyearbyen Airport were that in periods when Chinese tourists mostly came to Svalbard, Chinese nationals occupied 80% of the seats. However, accurate numbers are not available. In addition, an unfortunate issue with existing statistics is that they only cover information from official accommodation facilities in Longyearbyen. Expedition cruise ship trips are very popular among Chinese guests. Most of them never stay overnight in Longyearbyen, meaning they are invisible in the statistics, which again means that the actual volume is unknown and most probably much larger.

China itself is a country that has profoundly changed lately, and there are many ongoing processes that influence the who, when, and why of leisure travel outside China. Chinese tourists are changing, and we do not meet the same groups who visited five years ago. My participants widely agreed that Chinese travelling in groups are rather middle-aged or elderly, and they speak poor or no English. Further, they might be insufficiently informed beforehand regarding what kind of clothing, equipment, and skills they needed to visit Svalbard. One of my interviewees, a translator and guide for Chinese passengers, estimated that about 90% of the Chinese who come in summer travel in big groups and have a tour leader, who might but also might not speak English. The company responsible for the cruise has no control over that as the trip is booked through a travel agency in China. Cruises popular among Chinese guests are rather cheap, short, and calm, without operating in difficult waters. The clients are often 50+, and it is not rare to have people in their seventies on board. They come from all over China.

Yet the highly dynamic Chinese market is now moving to the new upper middle class, people born in the 1970s and later who work 9/9/6 (from 9 am to 9 pm 6 days a week), speak decent English, bring advanced equipment, and are willing to invest money in their offspring. This emerging clientele also comes during the period between October and January, traditionally seen as "low season".

Why Svalbard: the tight link between politics and representation

The Arctic is gaining the attention of Chinese officials, and it is impossible to fully separate geopolitics from tourism. It is too tempting to also add research. To illuminate this is the case of The Arctic Yellow River Station that was established by the Polar Research Institute of China in Ny-Ålesund in 2004 (Paglia, 2019). The station is a popular place to visit for Chinese tourists (for an anthropology of country-related tourist places see Kim et al. in this book). Tourism and research are the two new "soft extractive industries", and geopolitical motivations translate into both arenas. Modern China is building its image on highlighting the great past and the ever-greater future of the country. It is thus not accidental that Arctic destinations are becoming popular among Chinese visitors when China as a superpower aspires to engage more in the polar region(s).

In 2010, the world exhibition took place in Shanghai, and films made in Svalbard were presented there. Yet Svalbard is still not widely known in China, and those who come usually travel from another destination in the Arctic (if by boat) or Northern Scandinavia (if by plane). To visit the exotic and challenging Arctic means to confirm one's high social status (understood as *sensu,* Bourdieu, 1984), and the push toward unique experiences is strong. The mutual dependency between destination marketing and the guest's expectations is obvious. This dependency further influences how Svalbard is represented in media coverage, social media included.

Availability of light determines a lot in Svalbard, where the sun is seen for the last time by the end of October and first blesses Longyearbyen again by the beginning of March. Tourism products have to adapt to the rhythm of the seasons as well. My participants involved in the cruise ship industry operating around Svalbard (in June, July, and August only) agreed that Chinese visitors expect stunning observations of wildlife and also to get as far north as possible. "For them, especially, it is important to be able to come home and say, wow, look, I have just been at the 81st parallel North in the ice, and I saw polar bears" (Focus group, 5 May 2020). However, this is not only a driver for Chinese travellers. In a world where people and possibilities are many, and it gets ever more difficult to show off, exclusivity communicated through a simple selfie or a professional nature photograph posted on social media grants you a comparative advantage. "If they get these good selfies, they are happy and satisfied" (Interview, 9 April 2020). The problem here is that what is desired by many becomes also consumed by more and thus loses the label of exclusivity. In Svalbard, it also includes the painful burden of pressure on an ecosystem that needs protection. The visual aspect of marketing certainly plays the key role here:

> The more you send pictures like this to the Chinese market; the more Chinese will want to come and see it. It's a self-reinforcing effect, because if they come, they take their own selfies and post them on Viber or WeChat and then their neighbours and friends will want to come, too. We should avoid showing the same types of images or downplay the polar bear (Focus group, 5 May 2020).

In the confusing atmosphere of 2020, so heavily impacted by the global COVID-19 pandemic, there seems to be a univocal consensus of stakeholders declaring interest in local impacts of tourism, specifically, the quality of the experience and the learning aspect of the trip must be further pursued. Such goals poorly match speed and ephemerality, some of the key features of today's world. A receptionist in a hotel would comment:

> People come with exaggerated ideas about Svalbard. There are people from China who know nothing. Not even the temperatures. We have had so many Asian tourists this year asking where the boat to the North Pole is. Coming in sneakers. And then they are renting scooters. I am just wondering. Because it's so accessible. There are more and more flights, it's Instagram, people are posting pictures and everybody is so disappointed if they go for a trip and they don't see a polar bear. It's a disaster.

What the industry expects from Chinese visitors is the willingness to purchase. Shop and hotel owners in Longyearbyen appreciate their eagerness to spend money even though no statistics are available to confirm this widely shared perception. A typical quote here would be: "They like expensive goods. If they see a shop with cheap stuff, they won't enter". I also came across a statement pointing to the concept of bàn shǒu lǐ (伴手礼):

> There is a word in Chinese, something related to trust and networks. It's really important in Chinese culture to bring home a present. This present helps to maintain the relationships in your social network. That's why the Chinese tourists are such spenders.

The focus of this book is to explore Asian mobilities consumption, using Veblen's theory of conspicuous consumption as a starting point. As I could not talk to any Chinese visitors due to language incompetence combined with pandemic impacts, I chose not to focus primarily on their consumer behaviour. My perspective is contextualised by local perceptions of and experiences with Chinese tourism. As, for example, O'Cass & McEwan (2004) argue, conspicuous and status consumption are related yet not identical phenomena. Deepening the topic is beyond the scope of my chapter, but I hope other scholars can follow that thread.

So far, the picture seems somewhat superficial: wealthy Chinese who are privileged enough to be able to travel to the High Arctic come to fetch their pictures of vulnerable nature, shop, and leave. And indeed, there are four major problematic issues (if we ignore the deeper dilemma of intertwined geopolitical interests and the downside of global tourism in a warming world) that we need to address.

Scale and scope of marketing, communication, racism, and lack of strategy

To compare the destination of Svalbard (less that 3,000 inhabitants) with the People's Republic of China (over 1.4 billion inhabitants) means to imagine a dwarf and a giant. The levels of marketing budgets and exposure/outreach are incompatible. The potential demand in China is enormous, and it is difficult to gain an overview over the who, where, how, and why of Svalbard publications. As Svalbard is unique in terms of knowledge, competence, and equipment needed when exploring, there is a fear that much of the information being spread is misleading, and there is little capacity to influence that: "I don't think we have a chance", as a Visit Svalbard employee put it bluntly. In order to spread the information, you need to be visible in unknown channels, and in order to achieve the right reach-out, you need to pay a costly service to a specialised office based in Oslo or Copenhagen (which seems to be the capital of intermediary companies marketing Scandinavia in China). "I only know there is big money in the loop of marketing Svalbard, and this money is not always used in a way that we think is the right one in how to present Svalbard". Visit Svalbard spends about 1.3 million NOK a year on destination marketing, which is a rather scarce budget, but it is not likely that it will increase in the near future. "It's the others who are leading destination marketing for us. Our capability to influence the future is heavily limited". This problem is by no means a unique challenge for Svalbard, as other contributions to this book also demonstrate.

Marketing is closely related to media, including social media coverage. Many notice that there is a substantial interest in Chinese film crews coming to Svalbard. The perception is rather negative (being loud, pointing the camera into people's faces without asking for permission etc.).

> The difference between a Chinese crew and a crew from another country is that most of the others write to us beforehand and ask whether we can help them. "Can you help me to find somebody who I could talk to about the seed vault, or climate change?" The Chinese don't contact us. They just show up (Interview, 20 February 2020).

To achieve an improvement is not easy: "I believe it is important to prepare them before they arrive, but that's the challenge of China—we can't get to them before they come" (Interview, 20 February 2020).

The issue of inefficient communication is multi-layered: language (both as a means of communication and a symbolic system), culture (behaviour, customs, beliefs, understanding of leisure, privacy, concept of nature, etc.), and platforms (those used in Europe differ from tools used in China). The impossible task of communication results in stories about Chinese ladies with artificial diamonds on their eyelashes who destroyed a scooter driving without a driving licence on their way to the East Coast or groups of people with professional close-up lenses taking pictures of annoyed children and furious staff behind a kindergarten's fence.

The low level of cultural proximity and the trouble with distinguishing between individual Chinese visitors and China's political interests might count as factors underpinning racism.

It is now November 2019; I am sitting on a plane from Tromsø to Longyearbyen. One of my co-passengers engages in small talk and asks the obligatory questions that people in Svalbard often tease each other with: why are you travelling up there, what is your job, how long are you planning to stay. I feel a moral obligation to reply, as asking questions is my daily routine. The lady reveals she is a miner's girlfriend and spontaneously starts to comment on probable identities of the people on board. She is extremely critical towards the tourist industry, feels that the state betrayed the mining community with the decision to shut down the mines and open up the community to international tourism. All of a sudden, she makes a statement: "We, people who live in Longyearbyen, we hate tourists. And the Chinese, those are the worst of all".

It is not the first time I hear such a harsh comment. Local stakeholders who wish to work further with the Chinese know that precautious measures are needed. One of my participants from the guiding community told me that Chinese clients contacted him after being refused by a couple of tour operators: "There is no desire to understand them". He has a story to illustrate this. Last year, he had a group of 16 Chinese guests. Three guides were assigned to take care of them. There were two tour leaders and a six-year-old child in the group who spoke English. He felt he needed to protect them from racism. At the airport, they were refused when they asked to hire a car, and my participant had to tell a white lie to the tour leader to explain why he must drive the car himself. When booking a table for them in a restaurant, the waitress reacted with a "Yuck!" when he admitted the reservation was being made for a Chinese group. He chose a protective strategy and did his best to avoid letting the group interact with other people.

The three problems discussed above—scale and content of marketing, communication, and racism—all point in one direction, namely a missing common strategy. Visitors from other countries, especially Nordic guests, are preferred, and money is being invested in order to attract those kinds of tourists, as they are "easier to deal with" when it comes to language, equipment, and skills.

Other stakeholders, who have a broader scope, do have some strategic plans but nothing specific to Svalbard. Visit Norway is aiming to attract a more narrowly specified type of visitor. The goal should no longer be quantity but travellers who are likely to spend more time and money in the destination and who are "not having Norway as one of the eight destinations they should visit within one week" (Focus group, 5 May 2020). Younger couples or small families interested in the Scandinavian lifestyle and learning experience are perceived as promising.

These visitors also come outside the usual summer season, and they might combine their trip to Tromsø with Svalbard.

AECO has no specific strategy for the Chinese market, as they are not a commercial subject engaged in destination marketing.

> There are of course some of our members, for whom it is extremely important to do business with the Chinese market. Some members come with Chinese charters, that means that there are only Chinese people on board and there can sometimes be 180–200 Chinese people on a ship around Svalbard (Focus group, 5 May 2020).

An interesting stakeholder in this context is Hurtigruten Svalbard, the largest tour operator that has a sort of local connection. It is a daughter company of the much larger Hurtigruten with the national headquarters in Oslo and owners in Great Britain. Hurtigruten Svalbard has a seller in China with offices in several cities and allows for booking trips to Svalbard through Fliggy or Taobao.

When I presented my findings to Visit Svalbard, my conclusion about a missing strategy was considered wrong; the decision not to tackle challenges related to Chinese visitors is intentional, and it also reflects the current trend of drafting strategies according to target groups and not nationalities.

During my fieldwork, I met Robert, a representative of another company that has a thorough and innovative strategy for how to connect China and Svalbard. I believe it is useful to narrate Robert's story in more detail and do so in the next section.

Specialisation and nature education travel: the way to go?

Robert owns a company founded in Svalbard and a daughter company in China. There is no intermediary stakeholder in between. Robert works exclusively with Chinese guests, and Svalbard is one of his company's destinations. There are four stable employees, and they hire freelance guides. The company based in China is a fully owned foreign enterprise, which allows them to do business there, including bank transfers. Robert describes the market in locked down China under the pandemic as "very eager—first of all, they are all sitting with their mobile phones and they want to consume something". When travel is restricted, the time is optimal for marketing and reaching out. Robert identified a blind spot in the market that nobody was able or willing to exploit: the dark season. He observed that people in Longyearbyen have "socially and professionally accepted this drop in tourism numbers". The dark season is the time of social gatherings, wining and dining, and the Blues Festival. "But the hotels are at 20%, which is a disaster. And why is this? Because Norwegians normally like to do business with Norwegians and maybe Scandinavians". But Nordic people are unlikely to travel to Svalbard in that period. "We don't think it's fun. But if you look at the numbers coming from Asia in this time period—it's packed". As Tromsø became the capital of northern lights tourism, tourists visiting there all the way from Asia wish to experience something else in addition. Svalbard is then an ideal solution. Robert believes that local stakeholders in Longyearbyen have no ambition to reach out to those people. "The government? Do they really want an invasion of Chinese tourists? No. So why should they pay a lot of money to go and market?" He was convinced that it was possible to win that market easily. In order to succeed, though, he needed substantial intercultural competence.

Usual marketing tools are useless for the Chinese market; no search engine optimisation designed for Google will have an effect whatsoever, and nobody is on Instagram or Facebook. The biggest travel event is the national holidays period in October. Robert has a clear idea about how to present the concept to make it attractive to his market:

> We are Arctic specialists. We are not a travel company. We are not one in a million. In China, people say: "I am an expert. I have been to Longyearbyen, I am an expert". Okay. But we are specialists. We can do things that no one else can do.

The game changer is combining his expertise and Nordic look with a Chinese team behind him, which fosters trust.

> Because in China, the level of trust is very low. You don't trust people in China because you have all this copy culture. You call it "copy with pride". If I make an itinerary with pictures, someone will just copy it and try to sell it.

Robert illustrates the importance of cultural understanding with an anecdote. A potential client invited him to a tea tasting event. The host prepared the tea. He spoke no English but a lot of Chinese, for about three hours. Robert believes he made a good impression by being polite and patient. His humbleness made his client trust him enough to transfer money. "In our culture, when we make a deal, we just handshake and we trust each other. Not in China". It took Robert a year in China to acquire the necessary cultural competence.

There is a potential product suitable for connecting innovative trends in tourism to the unique opportunities of Svalbard. The new concept of nature education travel has been booming recently. "For Chinese kids, they have no idea about nature; they think walking in the park is nature". Interest is growing fast and demand is huge. Finland has been successful on this stage ("You learn how to walk in the forest, pick berries, build a fire—simple things"), but Norway is more prestigious than Finland. And what could be the most exclusive place in Norway if not Svalbard? "You need the product. But there are buyers all around China. They would be fighting to buy this. They believe education is everything. If the kid will learn something, they will pay anything for it". There are two points to consider when developing such a product: 100% safety and packaging "with an educational twist". Robert assured me that the clients would respect their "nature teachers" and behave with perfection. "In ten, 15 years, Chinese tourists with bad manners will fade away. But it's up to us to develop the programs".

Getting ready for an unpredictable future

The pandemic that hit China in December 2019 and Norway in March 2020 created a profound rupture in the recent Chinese tourism trend. How can Svalbard, with its tourism industry down on its knees, get ready for the highly insecure future when it comes to Chinese tourism?

None of my participants was expecting a decrease in interest when international travel is possible again, even if certain restrictions persist and become permanent.

> We are going to see a big increase in Chinese guests when the virus is gone. People usually book a trip at least a year in advance. And then they couldn't get out. What's gonna happen afterwards? They're gonna book again and they will try to come as soon as possible (Interview, 9 April 2020).

Neither Visit Svalbard, Visit Norway, nor AECO aims for increasing the volume and pressure, which is already elevated in terms of environment, infrastructure, and local community. The four problematic areas identified above can be tackled by developing a participative and inclusive local strategy that will take the following issues into consideration.

Communication is a key starting point. Without being able to exchange ideas and transfer knowledge, we risk seeing Chinese tourists coming to Svalbard as a large illegible threat. The issue has much in common with the inconvenience of having "too many" non-Norwegians settle down in Longyearbyen. The case of Svalbard illustrates the vulnerability of a small-scale travel destination exposed to an uncontrollable variety of marketing channels. It is easy to feel overwhelmed. Yet, there are measures that can be implemented, cost little, and help substantially. Many of the Chinese visitors coming to Svalbard have limited, incorrect, or no information about the environment and local manners. The reason is that "nobody tells them what is right and wrong, nobody tells them it is not permitted to walk in somebody's house". Often the tour leaders do not know either. As it is impossible to reach out to all charter companies and travel agencies selling Svalbard, the only bottle necks are Longyearbyen Airport and Port. A possible solution could be Community Guidelines (including polite habits while eating, taking pictures, or walking in town), translated into Mandarin, downloadable to smartphones through a QR-code on a visible banner. This might be combined with more advanced measures such as informative meetings in Chinese in the hotels when larger Chinese groups arrive.

Communication is also crucial when it comes to emergency issues, which differ in the summer and winter season. Language and cultural barriers add to the complexity in case of a major emergency event. Tackling the issue of emergency stands high on the agenda of Norwegian authorities responsible for the future development of Svalbard tourism (Ikonen & Sokolíčková, 2019), and measures will be taken soon. To what extent the tourist industry itself can contribute to better solutions, e.g. through financial support of improving emergency and health services available locally, something very much desired by the residents of Longyearbyen, is yet to be seen.

When it comes to destination marketing, it is necessary to distinguish between the different segments of the target group. While an increase in volume during summer is undesirable, Chinese (and Asian in general) tourists visiting during the dark season are an economically promising group that might help Svalbard tourism become an all-year-round business. If that is desirable, a more proactive approach by local stakeholders would be a logical step to take.

Better communication and optimalisation of marketing would contribute to "less painful meetings" between Chinese visitors and local residents. A positive factor would also be tailored products targeting the Chinese clientele, adapting activities to their needs. The level of adventure can be different compared to that which a Scandinavian guest seeks, and it is wise not to mix Chinese visitors with other nationalities. The unique concept of nature education travel targeting Chinese children and young adults can be further explored. While there is an obvious danger of making Svalbard available only to wealthy guests and missing other genuine guests interested in the destination, it might play a role in the country's further developments to which extent (young) Chinese citizens travel worldwide and get confronted with Otherness.

The potential of education can hardly be overestimated, especially combined with top Arctic science conducted in Svalbard by a flourishing international community of scientists studying issues related to climate change. Collaboration between science and tourism needs an interdisciplinary approach and the willingness of both managers of tourism and science, along with scientists themselves, to see where the potential lies. As one of my participants shared their perspective, "In China, they don't have the freedom to say anything they want. So, they don't have ideas. Our advantage is that we can go to China with a lot of ideas because we are allowed to have them".

Part of the future common local strategy should be mechanisms that prevent the development of so-called "domestic chains". These chains, known in other Arctic destinations such as Lapland and Northern Canada, include the purchase of land or real estate, charter flights, ownership of vessels, hotels and restaurants, importation of domestic employees, and even souvenirs and a portfolio of activities independent of local competence. These domestic chains make the local added value of (Chinese) tourism equal to zero; they contribute to an elevated pressure and feeling of alienation, and are thus to be avoided.

> It's only a matter of time until the Chinese copy the whole concept and run this business completely on their own. They are already working on it there. They have money to build the thing on a completely different scale than we are used to (Focus group, 5 May 2020).

To my knowledge, the only Chinese owner of property in Longyearbyen is currently Po Linn Lee, an entrepreneur from Hong Kong. The local newspaper, Svalbardposten, wrote about her activities in relation to the case of her

9.3-meter-high red Santa Claus mailbox that she had to remove at her own expense (Ylvisåker, 2016) as well as the large hotel project that she was planning to launch in the town's centre but eventually did not receive the necessary permissions from authorities in charge (Ylvisåker, 2017). Perceptions of Po Linn Lee's business in town are polarised; some feel she was discriminated against, while others watched her plans with worries. A potential Chinese investment in Hurtigruten Svalbard's mother company heavily impacted by the pandemic is locally and nationally feared (Totland, 2020). It is known that Chinese subjects have repeatedly expressed interest in buying land and property in Svalbard (The Barents Observer, 2016), which is much more than just an archipelago close to the North Pole. "The Arctic is the theme of the day in Beijing. So, we will be meeting the Chinese in the future for sure", as my AECO participant emphasised.

Tourism might rebound again, but the numbers are expected to reach the 2019 level no earlier than 2022. The involuntary pandemic break might be an excellent opportunity for developing a sustainable strategy for Chinese tourism in Svalbard.

Acknowledgements

The research was financed through the project CZ.02.2.69/0.0/0.0/18_070/0009476 (bo)REALIFE: Overheating in the high Arctic: Qualitative anthropological analysis funded by the Ministry of Education, Youth and Sports of the Czech Republic, and University of Hradec Králové.

References

Arlov, T. B. (2003). *Svalbards historie*. Trondheim: Tapir.

Bourdieu, P. (1984). *Distinction: A Social Critique of the Judgement of Taste*. London: Routledge & Kegan Paul.

Dvorak, Z. (2020). *Bærekraftig turisme i Arktis? En studie av utviklingen på Svalbard.* [Unpublished master's thesis]. Nord University.

Eriksen, T. H. (2015). *Small Places, Large Issues: An Introduction to Social and Cultural Anthropology*. London: Pluto Press.

Eriksen, T. H. (2016). *Overheating: An Anthropology of Accelerated Change*. London: Pluto Press.

Hindley, A., & Font, X. (2018). Values and motivations in tourist perceptions of last-chance tourism. *Tourism and Hospitality Research, 18*(1), 3–14.

Hovelsrud, G. K., Kaltenborn, B. P., & Olsen, J. (2020). Svalbard in transition: Adaptation to cross-scale changes in Longyearbyen. *The Polar Journal*, doi: 10.1080/2154896X.2020.1819016

Ikonen, E., & Sokolíčková, Z. (2019). *Optimal Tourism Balance Workshop: What are the research needs for knowledge-based tourism management in Svalbard?* Visit Svalbard and Association of Arctic Expedition Cruise Operators (AECO). https://en.visitsvalbard.com/dbimgs/OptimalBalanceWorkshop_Report.pdf

Kaltenborn, B. P., Østreng, W., & Hovelsrud, G. K. (2020). Change will be the constant—future environmental policy and governance challenges in Svalbard. *Polar Geography, 43*(1), 25–45.

Kelman, I., Rauken, T., & Hovelsrud, G. (2012). Local business perceptions of weather impacts on tourism in Svalbard, Norway. *Northern Review*, (35), 96.

O'Cass, A., & McEwen, H. (2004). Exploring consumer status and conspicuous consumption. *Journal of Consumer Behaviour*, *4*(1), 25–39.

Paglia, E. (2019). A higher level of civilisation? The transformation of Ny-Ålesund from Arctic coalmining settlement in Svalbard to global environmental knowledge center at 79° North. *Polar Record*, *56*, 1–13.

Saville, S. (2019a, August 30). The Northern-most Overtourism? [Conference presentation]. *Royal Geographical Society with IBG Annual Conference*, London, United Kingdom.

Saville, S. (2019b). Tourists and researcher identities: Critical considerations of collisions, collaborations and confluences in Svalbard. *Journal of Sustainable Tourism*, *27*(4), 573–589.

Sokolíčková, Z. (2020, September 9). *Turisme, Svalbard og Kina. Svalbardposten.* https://svalbardposten.no/turisme-svalbard-og-kina/19.12916

Staalesen, A. (2016, October 21). No Chinese resort after all. *The Barents Observer*. https://thebarentsobserver.com/en/arctic/2016/10/no-chinese-resort-svalbard-after-all

Totland, P. A. (2020, October 2). Slik kan Kina feste grepet om Svalbard. *Svalbardposten.* https://svalbardposten.no/kronikker/slik-kan-kina-feste-grepet-pa-svalbard/19.13028

Van Bets, L. K.J., Lamers, M. A.J. & van Tatenhove, J. P.M. (2017). Collective self-governance in a marine community: Expedition cruise tourism at Svalbard. *Journal of Sustainable Tourism*, *25*(11): 1583–1599.

Viken, A. (2006). Svalbard, Norway. In G. Baldacchino (ed.). *Extreme Tourism. Lessons from the World's Cold Water Islands* (pp. 129–142). New York: Elsevier Science.

Wu, Hung-Che, & Chang, Ya-Yuan. (2020). Arctic attachment and experiential relationship marketing in Arctic tourism. *Scandinavian Journal of Hospitality and Tourism*: *Asia and Arctic Tourism*, *20*(2), 178–203.

Xie, J. & Tveterås, S. (2020). Economic drivers for the Chinese tourists. *Scandinavian Journal of Hospitality and Tourism*, *20*(2), 110–125.

Ylvisåker, L. N. (2016, September 9). Ikke mer julepost. *Svalbardposten.* https://svalbardposten.no/nyheter/ikke-mer-julepost/19.7534

Ylvisåker, L. N. (2017, July 21). Fra hotell til studentboliger. *Svalbardposten.* https://svalbardposten.no/nyheter/fra-hotell-til-studentboliger/19.8714

12 Embracing Chinese tourists in Iceland

The response of the tourism industry

Vera Vilhjálmsdóttir, Guðrún Þóra Gunnarsdóttir, and Rosemary Black

Introduction

The tourism industry in Iceland has seen significant growth in the past decade, from nearly 500,000 tourists in 2010 to two million in 2019 (Ferðamálastofa, n.d.). The tourism boom has brought both opportunities and challenges for the industry, authorities, and communities around the country (Sæþórsdóttir et al., 2020). With the growing numbers of tourists, the diversity of the people visiting Iceland has increased.

The most prominent new tourism market is from China, growing from 5,000 to 99,000 visitors in one decade (see Figure 12.1). Since June 2017, departing passengers from Hong Kong and Taiwan have been recorded specifically. In 2018 and 2019, approximately 70% of Chinese tourists came from Mainland China, 20% from Taiwan, and 10% from Hong Kong (Ferðamálastofa, n.d.).

Chinese tourists visiting Iceland are well educated, and 80% have average or above average income levels. Almost half (48%) are 25–34 years old, and 64% are women (Oddný Þóra Óladóttir, 2019). This is consistent with the demographic profile of Chinese tourists visiting other Nordic countries and Europe (The Nordics, 2018; World Tourism Organization (UNWTO), 2019).

According to a year-round visitor survey carried out in 2018 by the Icelandic Tourist Board and Statistics Iceland, Chinese tourists cite "Icelandic nature" (99%), "wanting to try something new" (94%), and an "interest in the Arctic" (88%) as three main reasons for visiting (Oddný Þóra Óladóttir, 2019). Icelandic nature is the main attraction for all international tourists; however, Chinese tourists score marginally higher in this regard. Interestingly, Chinese tourists are more influenced by media coverage than tourists of other nationalities. Thus, 88% of Chinese tourists (versus 50% tourists of other nationalities) stated that "coverage about Iceland online" influenced their decision to visit Iceland. Additionally, 80% of Chinese tourists versus 37% tourists of other nationalities indicated that "Icelandic nature in films and TV" affected their decision making. Furthermore, the image of Iceland as a safe travel destination is an important influencer in Chinese tourists' decision making (ibid).

Finally, Chinese tourists have increasingly become an important part of winter tourism in the last few years. Two major one-week holidays in the Chinese calendar

DOI: 10.4324/9781003039518-12

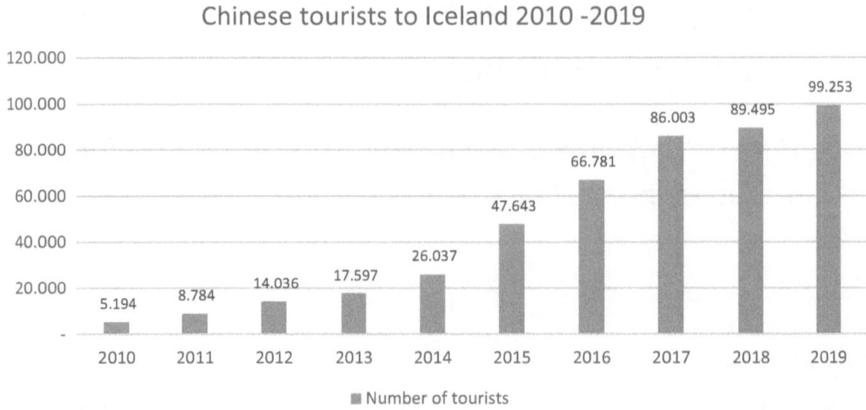

Figure 12.1 Number of Chinese tourists in Iceland 2010–2019. Source: (Ferðamálastofa, n.d.)

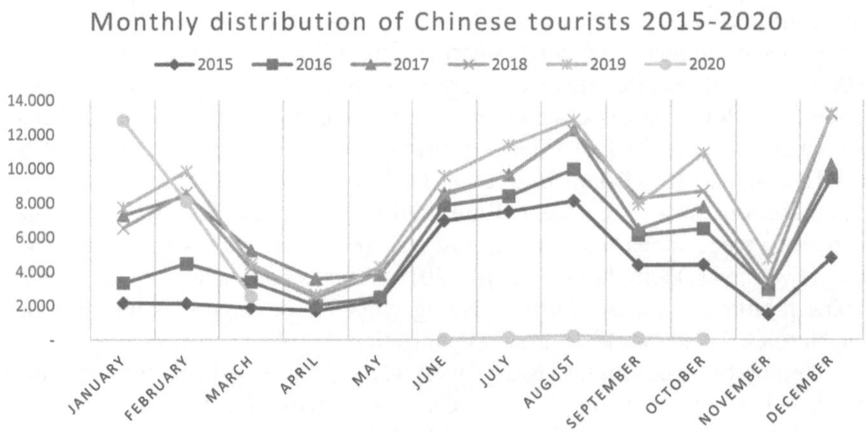

Figure 12.2 Chinese tourists' arrivals by month 2015–2020. Source: (Ferðamálastofa, n.d.)

year fall during the winter season, one in October and the other usually in February (World Tourism Organization (UNWTO), 2019). Around 49% of Chinese tourists named the Northern Lights as the main attraction of Icelandic nature (Oddný Þóra Óladóttir, 2019). A wide range of Northern Lights tours are on offer at varying price points (Heimtun et al., 2014), and this has helped make Iceland an attractive destination for Chinese tourists during the winter (see Figure 12.2).

In recent years, Icelandic tourism authorities, along with the industry, have emphasised the need to address seasonality to create more year-round job opportunities (Árnason & Welling, 2019). Hence, the increased arrival of Chinese tourists during the winter months has been important and is a trend that seems to be becoming stronger each year. In January 2020, there was a significant increase from the previous year in Chinese tourist arrivals (see Figure 12.2) before the COVID pandemic occurred.

This chapter aims to shed light on how the Icelandic tourism industry has experienced hosting and interacting with the growing numbers of Chinese tourists. First, we provide a brief summary of the development of China's outbound tourism, as well as citing relevant research on how other Nordic and Arctic communities have addressed the arrival of this new market. Following a methodology section, the results of 12 interviews with various tourism industry representatives are presented and discussed. Finally, we end with concluding remarks and recommendations.

Chinese tourists in the global, Nordic, and Icelandic context

Chinese outbound tourism has grown exponentially in the 21st century, from 4.5 million outbound tourists in 2000 to 150 million in 2018. It has been the largest source market for tourism in the world since 2012. In addition to the opening up of the Chinese borders, other factors such as a growing middle-class, more disposable income, and greater purchasing power have all contributed to increased travel by Chinese tourists (The Nordics, 2018; World Tourism Organization (UNWTO), 2019). Traditionally, group packaged tours for overseas travel have been very popular with the Chinese, mainly due to the convenience they present, the limiting of language barriers, and their reasonable pricing (Chen et al., 2018). Studies indicate that younger Chinese tourists are more likely than the older generations to travel on short notice and travel independently (The Nordics, 2018). When it comes to choosing their destination, beautiful or scenic tourist attractions are the main factor in their decision making (38%), followed by pricing (16%) and activities available (15%) (World Tourism Organization (UNWTO), 2019). Destinations in other Asian countries are the most popular among Chinese tourists, for example, short haul trips to Thailand, Japan, South Korea, and Vietnam. However, long haul travel has become more frequent, particularly in relation to an increased interest in the Arctic as a destination in recent years (Lee, 2020; World Tourism Organization (UNWTO), 2019). The growing interest in the Arctic can be attributed to factors such as globalisation and climate change. Due to the image often given of the Arctic region as a vulnerable and endangered region, tourists are putting it on their bucket list as a "last chance" destination (Carson, 2020).

The Nordic countries have increasingly been paying more attention to the Chinese market and have focused on becoming "China ready" in the past decade (Jørgensen & Ren, 2015; Lee, 2020). Thus, the Nordic DMOs have collaborated in an effort to increase the Nordic countries' visibility to the Chinese market (The Nordics, 2018). Each of the Nordic countries is at different stages in this process and face varied challenges regarding this market. For example, while search terms relating to Iceland Travel have the highest volume of any Nordic country, Iceland has no official online platforms hosted in China (The Nordics, 2018).

Visit Iceland's website has information available in Chinese, but since it is not hosted in China, the page loads much slower than Chinese users are used to due to China's internet restrictions (ibid). Likewise, Visit Greenland has faced difficulty in how to market the country as a travel destination for the Chinese market and ensure that Chinese travel agents understand the realities of Greenland's travel conditions (Martin & Ren, 2020).

Chinese tourists often combine two or more Nordic countries in the same trip (The Nordics, 2018; Wonderful Copenhagen, 2017), and Iceland has increasingly become a gateway destination to Greenland and other more northern destinations (Jóhannesson & Welling, 2020). Scandinavian tourism practitioners have commented that Scandinavia is less likely to be the first choice of destinations for Chinese tourists but as a destination is more likely to attract more seasoned tourists (Jørgensen et al., 2018). Furthermore, the Nordic countries are not as familiar destinations to Chinese tourists as other destinations in Europe. They, therefore, seem to have less preconceived expectations, and are open to new experiences and to see something unexpected (ibid).

To date, the focus of research in this field has been on the demographics of Chinese tourists visiting the Nordic countries and their travel behaviour and satisfaction, and much less attention has been on the experiences of the Nordic countries hosting Chinese tourists. Apart from one master's thesis (Xing, 2020), no studies have been conducted in Iceland on issues relating to this recent growth of Chinese tourists' visitation to Iceland. This study seeks to fill a research gap in the knowledge about Chinese tourists in Iceland and answer the research question of how tourism practitioners in Iceland have experienced and responded to the recent upsurge in Chinese tourist arrivals.

Methodology

This qualitative study explores the Icelandic tourism industry's experiences of hosting Chinese tourists. By using purposeful sampling (Maxwell, 2013), a spectrum of industry insiders from various facets of Icelandic tourism were contacted. In particular, emphasis was put on interviewing informants who a) had experiences of face-to-face interactions with Chinese tourists, b) had been engaged in the tourism industry for a significant period of time, and/or c) had extensive knowledge of the Chinese tourism market and its development in Iceland. Utilising these criteria, informants working as tour guides or tour operators, managers of hotels or car rental businesses, as well as a representative from Promote Iceland were contacted and interviewed. Promote Iceland is a public-private partnership that aims to increase the visibility and competitiveness of Icelandic businesses in foreign markets. Interview guides were created for each profession, although there was a set of core questions across all the interviews, such as background questions about the informants, the services they provide, and their experiences/challenges providing those services to Chinese tourists. The University of Iceland's Confucius Institute Northern Lights was consulted during the preparation of this project and in the creation of the interview guide.

Table 12.1 Overview of interviewees

Interviewee	*Gender*	*Experience and additional details of interviewees*
1	Male	Tour guide and tour operator, guides in Chinese, has worked for various tourism companies in Iceland and independently.
2	Female	Tour guide and driver, guides in English, has worked for various tourism companies in Iceland and independently.
3	Female	Tour guide, driver, and tour operator, guides in Chinese, of Chinese descent, business owner.
4	Female	Tour guide, guides in English, has worked for various tourism companies in Iceland.
5	Male	Tour operator, speaks Chinese, has experience marketing Iceland in China.
6	Male	Hotel manager for a large independent hotel in South Iceland
7	Male	Chief booking manager for a large hotel in the Capital region, the hotel is a part of a larger Icelandic hotel chain.
8	Female	Chief operations manager for an Icelandic hotel chain in the Capital region.
9	Male	Station manager for a car rental company at Keflavík, the international airport.
10	Both Male	CEO and Chief service and HR manager for a large car rental company.
11	Male	Project manager at Promote Iceland/Visit Iceland, trade & invest, with a focus on the Russian and Asian markets.
12	Female	Former flight attendant for an Icelandic domestic airline, recently took over as hotel manager for a hotel in North Iceland.

In total, 12 semi-structured interviews (Crang & Cook, 2007) with 13 Icelandic informants were conducted from March to September 2020. The interviews were between 30–90 minutes long and all conducted in Icelandic. Due to COVID-19 related issues, the interviews were either telephone or video calls. The interviewees had on average 20 years of industry experience, ranging from five to 33 years (see Table 12.1).

The interviews were transcribed and the transcripts manually analysed and coded to identify the main themes, ideas, and concepts (Taylor et al., 2016).

Results

The overarching themes that emerged during the analysis of the interviews were the challenges relating to cultural differences and language barriers between the Chinese tourists and the Icelandic employees. These challenges manifest in issues such as safety concerns, especially during the winter, and finding the most appropriate ways to communicate important information about Iceland and Icelandic nature to Chinese tourists. Many of our interviewees talked about their learning curve once the number of Chinese tourists began to increase, as it took them time to understand the needs of this new group of tourists and how to best guide and communicate important information to them.

The interviewees emphasised the importance of understanding and appreciating the cultural differences to facilitate positive interactions between the Chinese tourists and themselves.

> Because [Chinese tourists] come from a different culture, where there are other values and norms, [service providers in Iceland] just need more information. [...] Because as soon as you understand people a little bit better, where they come from and what context they come from, then it becomes much easier to provide them with good service. (Interviewee 5)

Better insight into the Chinese culture and social norms was repeatedly mentioned, in particular by those service providers who spoke no Chinese. Some of the interviewees rendered stories of how their lack of understanding affected their service and their capacity to fulfil the needs of their customers "unfortunately I didn't know anything about the Chinese tourists and what I tried to use didn't work at all" (Interviewee 4).

Language issues

Language barriers appear to further complicate the cultural differences. Although communication difficulties are not limited to Chinese tourists, some interviewees mentioned the lack of English proficiency had been somewhat of a surprise, especially regarding younger Chinese travellers. Chinese tourists travelling in groups often speak little English, so hotel staff mainly communicate with the tour guides accompanying the groups.

> In big groups, the individuals don't talk to the staff much, they don't speak English. If there is something that needs addressing it is the tour guide that brings the messages [to the staff]. (Interviewee 7)

Other than keeping a good line of communication open to the tour guides accompanying the groups, the service providers have found different ways to deal with the communication barriers. One is to make sure that the staff go over all the information needed in English with independent travellers, "everything gets across if you speak slowly, they know 'breakfast' and they know 'restaurant' and 'Wi-Fi'" (Interviewee 6). Hiring Chinese speaking staff and having signs and information translated into Chinese in the rooms and dining areas has also helped. Icelandic tour guides guiding in English have tried to find their own ways to connect with their groups by, for example, sharing pictures of grandchildren with older Chinese tourists (Interviewee 2) or entertaining younger tourists by acting "goofy" (Interviewee 4).

Car rental companies have also had to learn how to approach this new group of tourists and find the best ways to inform their Chinese customers about Icelandic road conditions and safe driving guidelines. The car rental company representatives stated that when Chinese FITs (Free Independent Travellers) started to visit Iceland and went on self-drive tours, they had on average more accidents than most other groups of tourists (Gunnarsson et al., 2020). To combat this situation, one of the car rental companies interviewed worked strategically to find solutions on how to get information across to their Chinese customers.

> We learned quickly and were told that [Chinese tourists] want to get the information before they come [to Iceland], because they aren't interested in having to read a lot once they've arrived, as then their holiday has started. (Interviewee 10)

By working closely with the Chinese embassy in Iceland and their international affiliated partners, they made and distributed an information booklet in Chinese and advertisements on Chinese online media sites. These and other preventive measures have resulted in fewer accidents and car damage by Chinese customers. Despite these measures taken by car rental businesses, other interviewees felt that increased winter tourism resulted in various challenges, and Chinese tourists needed to be better informed about road conditions, distances between destinations, and general road safety.

Out and about in Iceland: safety issues and travel itineraries

The Icelandic tour guides were generally concerned that Chinese tourists were not properly informed of winter conditions, and Chinese tour organisers too often used the same travel itineraries the whole year around. Travel plans that are not realistic or achievable can create friction and cause difficulties once the tour is underway. One tour guide, who has specialised in guiding and hosting Chinese speaking tourists, said she always insists on seeing the travel itineraries beforehand to rearrange them, so they better reflect the seasonal conditions (Interviewee 3).

The guides also pointed out that the Chinese tourists often wanted to see as much as possible during their visit and did not always appreciate the distances between attractions and that road conditions, particularly during wintertime, might prevent them from visiting all their desired places. This has sometimes resulted in what the guides labelled as unfair or unrealistic demands. In particular, demands outside the regular itinerary were mentioned "all of a sudden they realise that they are missing out on the glacier lagoon and start demanding that the driver drives them there" (Interviewee 1), as well as demands outside regular working hours:

> it was overcast in Gerði and they came with the weather forecast and said 'there are clear skies here' and that was by Eyjafjöll [...] it doesn't make sense to drive four hours in the middle of the night to some clear sky. (Interviewee 4)

This behaviour is especially prevalent in relation to Chinese tourists' eagerness to see the Northern Lights, as their disappointment in missing out on this experience can pose difficulties. Some Chinese tourists have asked for refunds, despite it being clear companies cannot guarantee Northern Lights in their tours (Interviewee 4). One tour guide thought the strong emphasis on seeing the Northern Lights created unrealistic expectations. When she gets questions about visiting Iceland and the client expresses an interest in the Northern Lights, she indicates she will try to show them to them, but Iceland has much more to offer and tells customers they should foremost come to see the country itself (Interviewee 3).

Adaption to service challenges

In general, the Chinese tourists were described by the interviewees as being polite and agreeable customers. However, certain challenges have arisen over the years that hotels and other tourism service providers have had to address and find solutions.

The hotel representatives said they had made different arrangements regarding breakfast when they have Chinese groups or have had to adjust to different breakfast habits. Chinese tourists like to eat a hearty breakfast, "we need to have more breakfast ready for them and we have not been saying to them that breakfast is over" (Interviewee 6) and some hotels have even started offering Asian breakfast items. Similarly, room cleaning has to be adjusted "we know [Chinese tourists] leave the rooms in a worse condition and we need to plan the cleaning accordingly" (Interviewee 6), "they don't damage things or anything like that, but their cleanliness could be better" (Interviewee 7) and "you sense more that there has been a guest in the room" (Interviewee 8). The interviewees described this more as a challenge that needed to be worked around, rather than a problem.

Some of the interviewees commented they had sensed a social hierarchy pertaining to social interactions between the Chinese and tourism providers. In particular, this manifested in the relationship between customers and service providers:

> I won't say that [Chinese tourists] look down on people in service jobs, but there is still [pause] because you are paying, a gap is created between you and the service provider. You are not required to be polite to this service provider because you are paying for a service [...]. So Icelandic people and people in the tourism industry experience that they are rude and demanding. (Interview 5)

This social hierarchy is not something Icelandic tourism service providers are accustomed to and they find it can manifest in unexpected ways.

> It was this hierarchy. Especially when they are travelling in a group, then it wasn't appreciated if our staff was talking to everyone. There wasn't an understanding [on our behalf] that it might not be appropriate for us to speak to someone else before we spoke to the leader. (Interviewee 8)

This social hierarchy also manifests in interactions with guides and other service providers having trouble getting Chinese tourists to follow the rules they set out for them, as the information is "just" coming from an Icelandic tour guide or a flight attendant (Interviewees 4 and 12). For Chinese tourists, it seems to make a difference from whom the information and communication is coming. They trust the government and are more likely to follow official state rules "after we consulted with the embassy [we learned] everything that's called a 'ban' is good to use on them. Just ban it, it's a police matter" (Interviewee 10). Instructions coming from the government or public institutions about what is and is not allowed in Iceland is more effective, than coming from tour guides or tour operators (Interviewee 3).

Engaging with Icelandic nature and culture

Most Chinese tourists who come to Iceland are from, and are used to, urban areas and human-influenced environments and rarely get to experience untouched nature and expansive landscapes as offered in Iceland.

Their reaction to being in Icelandic nature can therefore be different to the average international tourist, and tour guides describe unusual responses from the Chinese tourists to some tourist attractions. "I have been asked how many slaves carved out [the cliffs at] Reynisfjara" (Interviewee 2). The tour guide concludes that maybe it is not such a strange question if you look at the form of the basalt columns. This unfamiliarity with "wild" nature also means it is important to cover safety precautions at the natural attractions.

> At all stops we need to warn them explicitly about everything because there isn't fencing or other safety measures, there isn't anything like that at many of our places and then it's like they can't sense the danger. (Interviewee 4)

This different reaction to Icelandic nature does not only apply to safety concerns, but the tour guides also notice this extends to how the Chinese tourists approach nature.

> They are observing in their own way [...] you see them go out [of the bus] and run in a complete different direction to where the waterfall or geyser is, and start taking pictures of the flowers, the vastness, and the horizon. (Interviewee 1)

Tour guides try to stay close to the group when exploring, both to ensure their safety and that there isn't any doubt about what [the place] is about" (Interviewee 1) and answer any questions. One tour guide likes taking the opportunity to encourage her groups to experience nature in different ways, for example, by having fun in the snow with snowball fights and taking a moment to appreciate the silence in nature because "they have never heard [silence] before" (Interviewee 2).

Due to the language barriers, Chinese tourists may not always ask many questions, but when they do, the guides describe them as very curious about

Icelandic nature and they like hearing about the flora, fauna, and the geological history, but also about Icelandic society. They are interested in learning about contemporary issues and cultural norms in Iceland; "They are interested in things they can mirror themselves in or compare themselves to" (Interviewee 3). Tour guides often get questions about the cost of living in Iceland, wedding traditions, family structures, and the welfare system. Due to Chinese tourists' interest in the local culture, there are many unexplored opportunities for Icelandic tourism businesses to give Chinese tourists a glimpse into local life, "there are opportunities in, for example, inviting people home for a home cooked meal" (Interviewee 5).

Discussion

The Icelandic tourism industry has experienced significant growth in the past decade with the Chinese market increasingly becoming an important group, yet little is known about how the industry is responding and adapting to this new market. This study focused on the views and perceptions of select groups of industry representatives who have extensive experience hosting and servicing Chinese tourists. Our interviewees agreed that there are opportunities for the Icelandic tourism industry in the Chinese market. However, to ensure Chinese tourists have a great experience in Iceland, it is important to continue to improve the lines of communication with them.

Making official Icelandic tourism sites or profiles more visible on Chinese online sources is important to ensure information about travel in Iceland is being communicated accurately to ensure tourists arrive with realistic expectations. This is especially important considering the increasing number of Chinese FITs organising their own trips and doing self-drives, plus the increased importance of winter tourism. The conditions and remoteness of Arctic destinations can be difficult for tourists to navigate, especially during the winter months, as the winter conditions can prove dangerous due to factors such as snow and icing and can lead to increased risks of injury (Lépy et al., 2016). How to best inform tourists who are unfamiliar with Arctic conditions is something Arctic destinations need to take into consideration. The difficulties in conveying Nordic/Arctic realities to Chinese tourists are not confined to Iceland, as Greenland also faces similar challenges. Nature, distances between places, and general conditions of remote areas provide unique challenges that Chinese travel agencies have yet to fully grasp. This has led local tourism practitioners in both countries having to adjust itineraries developed by Chinese tourism operators (Martin & Ren, 2020). What this study further indicates is that different strategies for the dissemination of information about Arctic conditions is needed for certain groups, such as Chinese tourists, to ensure tourists are as prepared as possible before they reach the destinations.

Hosting and providing services to Chinese tourists was regarded as challenging by many of our interviewees, mainly due to perceived cultural differences. However, almost all interviewees described it as a learning curve and a matter of understanding Chinese culture and Chinese tourists' expectations. For example, while the Western world often emphasises the divide between nature and culture, East Asian culture views them as a unity, and this is one of the main tenets of East Asian philosophies (Lee et al., 2012). This means that how they react to and interact with nature can be different to Western tourists. Therefore, a better understanding of how Chinese people approach nature could help tour guides find interesting and safe ways for their Chinese tourists to engage with Icelandic nature in a way that makes sense to both parties. A learning process like this can lead to increased capacity and global skills in Arctic communities (Jørgensen & Bertelsen, 2020). In general, by understanding some core elements of Chinese culture, the Icelandic tourism industry could find ways to accommodate their wishes into their existing business models and anticipate and manage Chinese customers' expectations to reduce disappointment and social friction.

Conclusion

The results of this study have practical significance for the Icelandic tourism and hospitality industries. First, industry bodies and businesses need to develop and deliver cultural awareness staff training and basic Chinese language training for front line staff. The strategies implemented by the car hire companies demonstrate the value of tailoring advice and information for Chinese tourists. Second, given the specific interests and needs of the Chinese market, tour operators and tour guides may need to tailor tour itineraries to meet their specific needs. Further cooperation and exchange of information with Chinese tour operators licenced to bring Chinese tourists to Iceland would be beneficial. Third, with the increase in Chinese tourism during winter, it is particularly important to ensure that all marketing and promotional material delivered to the Chinese market prior to visiting Iceland is realistic and clearly outlines the challenges of visiting Iceland, such as the weather, distances, and topography. Finally, tourism and hospitality tertiary education providers need to include course content on the specific tourist groups visiting Iceland.

This study represents the first in-depth study into how the Icelandic tourism and hospitality industry is reacting to the increase in Chinese tourists' arrivals and provides some important suggestions for the industry to improve its response to this market that will result in an enhanced tourist experience. However, this study had a limited number of tourism and hospitality industry representatives. There are opportunities for further research into this topic, such as a larger qualitative study with more participants and covering an even wider cross-section of the industry, a national survey, and/or more focused studies of specific sectors of the industry.

Acknowledgements

The authors would like to thank the interviewees for their invaluable contributions and willingness to participate in the interviews. We also want to thank The Confucius Institution Northern Lights for their input and advice in the preparation of this project.

References

Árnason, Þ., & Welling, J. T. (2019). Winter tourism and seasonality in Iceland. In U. Pröbstl-Haider (ed.), *Winter Tourism: Trends and Challenges* (pp. 422–460). Oxfordshire: CAB International.

Carson, D. A. (2020). Urban tourism in the Arctic: A framework for comparison. In *Arctic Tourism in Times of Change: Dimensions of Urban Tourism* (pp. 6–17). Copenhagen: Nordic Council of Ministers.

Chen, N., Masiero, L., & Hsu, C. H. C. (2018). Chinese outbound tourist preferences for all-inclusive group package tours: A latent class choice model. *Journal of Travel Research*, 56(6), 916–931.

Crang, M., & Cook, I. (2007). *Doing Ethnographies* London: SAGE Publications Ltd.

Ferðamálastofa. (n.d.). Ferðamenn um Keflavíkurflugvöll. *Ferðamálastofa*. Retrieved November 20, 2020, from https://www.ferdamalastofa.is/is/tolur-og-utgafur/fjoldi-ferdamanna/talningar-ferdamalastofu-i-flugstod-leifs-eirikssonar

Gunnarsson, G. G., Þorsteinsdóttir, K. B., & Jónsdóttir, Þ. (2020). Umferðarslys á Íslandi 2019. *Samgöngustofa*. https://www.samgongustofa.is/media/umferd/skyrslur/Slysaskyrsla2019.pdf

Heimtun, B., Jóhannesson, G. Þ., & Tuulentie, S. (2014). *Northern Lights Tourism in Iceland, Norway and Finland*. UiT, The Arctic University of Norway. https://septentrio.uit.no/index.php/SapReps/article/view/3266

Jóhannesson, G. Þ., & Welling, J. T. (2020). Reykjavík as Arctic destination: Creativity and Urban Tourism. In *Arctic Tourism in Times of Change: Dimensions of Urban Tourism* (pp. 51–57). Copenhagen: Nordic Council of Ministers.

Jørgensen, M. T., & Bertelsen, R. G. (2020). Chinese tourism in the Nordic Arctic—Opportunities beyond the economic. *Scandinavian Journal of Hospitality and Tourism*, 20(2), 166–177.

Jørgensen, M. T., Law, R., & King, B. E. (2018). Beyond the stereotypes: Opportunities in China inbound tourism for second-tier European destinations. *International Journal of Tourism Research*, 20(4), 488–497.

Jørgensen, M. T., & Ren, C. (2015). Getting 'China Ready'. Challenging static and practice based configurations of the Chinese tourist. *AEI-Insights: An International Journal of Asia-Europe Relations*, 1(1), 19–34.

Lee, Y.-S. (2020). Asia and Arctic tourism. *Scandinavian Journal of Hospitality and Tourism*, 20(2), 105–109.

Lee, Y.-S., Lawton, L. J., & Weaver, D. B. (2012). Evidence for a South Korean model of ecotourism. *Journal of Travel Research*, 52(4), 520–533.

Lépy, É., Rantala, S., Huusko, A., Nieminen, P., Hippi, M., & Rautio, A. (2016). Role of winter weather conditions and slipperiness on tourists' accidents in Finland. *International Journal of Environmental Research and Public Health*, 13(8), 822.

Martin, N., & Ren, C. (2020). Adventurous Arctic encounters? Exploring Chinese adventure tourism. *Scandinavian Journal of Hospitality and Tourism*, 20(2), 126–143.

Maxwell, J. A. (2013). *Qualitative Research Design: An Interactive Approach* (3rd ed.). Thousand Oaks: SAGE Publications.

Oddný Þóra Óladóttir. (2019). Erlendir ferðamenn á Íslandi 2018: Lýðfræði, ferðahegðun og viðhorf. *Ferðamálastofa*. https://www.ferdamalastofa.is/static/files/konnun2018/konnun-18-2.pdf

Sæþórsdóttir, A. D., Hall, C. M., & Wendt, M. (2020). From boiling to frozen? The rise and fall of international tourism to Iceland in the era of overtourism. *Environments*, 7(8), 59.

Taylor, S. J., Bogdan, R., & DeVault, M. L. (2016). *Introduction to Qualitative Research Methods* (4th ed.). New Jersey: Wiley.

The Nordics. (2018). *Chinese Market Segment: High Spending FITs. Opportunities and Challenges for Nordic Destinations*. Copenhagen: The Nordics.

Wonderful Copenhagen. (2017). *Chinese Tourists are all kinds of Travellers: 2016 Survey of Chinese Tourists in the Nordic Cities.* Wonderful Copenhagen. https://www.wonderfulcopenhagen.com/chinavia/research/research-reports

World Tourism Organization (UNWTO). (2019). *Guidelines for the Success in the Chinese Outbound Tourism Market*. Madrid: World Tourism Organization.

Xing, Y. (2020). *Chinese Tourists in Iceland: Motivation and Satisfaction* [Master's thesis, University of Iceland]. Skemman. http://hdl.handle.net/1946/35798

13 Designing winter cities

Arctic urbanisation and Asian mobilities consumption

David Chapman, Young-Sook Lee, and Agneta Larsson

Arctic urbanisation and spatial planning

The Arctic is urbanising (Stokkan & Reiter, 1994; Varentsov et al., 2018). Settlements are growing in population and size, and their hinterlands are gaining more importance as centres for resource extraction and processing (Morata et al., 2020). Concurrently, Asian tourists to the Arctic have been increasing in recent years, particularly since 2010 (Jørgensen & Bertelsen, 2020). Asian tourists may well hold their own ontologically-based concept of "urban" while visiting Arctic spaces. Indeed, the past four decades have witnessed unprecedented demographic changes in Asian areas—inclusive of the Northeast, Southeast, South, and Central urban populations, which marked nearly 1.6 billion in 2010 (Pellegrini, 2011). A notable trend in Asian compared to Western urbanisation is its speed. Dating from the industrial revolution, Western urbanisation primarily increased over two and a half centuries. On the other hand, Asian urbanisation has intensified within the last century and contributed to many changes in people's perceptions of space, infrastructure, and more in the urbanising of Asian cities (Parai & Dutt, 1994; Zhang & Deng, 2017). This is where challenges to designing Arctic urban space may arise, particularly when attempts are made to incorporate Asian tourist needs and desires in the form of Asian mobilities consumption of space and culture in processes of urbanising the Arctic. Asian tourists' ways of consuming space may also have implications in other areas of consumption during their visit.

A key challenge in the rapidly urbanising Arctic is how to create urban space that is attractive to Asian tourists and their consumption. Here the opportunities for growth in tourism are tensioned with respect to pressures for urbanisation and resource extraction and the benefits this can bring to communities in terms of employment and economic growth.

Arctic settlements can be defined as places that experience temperatures below 0°C, precipitation mainly in the form of snow, and limited hours of sunshine and daylight, with these conditions persisting for up to six months of the year (Mathisen et al., 2015). These settlements also exhibit appreciable seasonal climatic variation (Pressman, 2004) and are small in size and scale in comparison to Asian cities. Arctic cities and their cold, snowy, and dark winters are occasionally lit up by the Northern Lights in the night sky, making them attractive and magical

DOI: 10.4324/9781003039518-13

destinations in addition to creating unique tourism experiences (Jóhannesson & Lund, 2017). This unique experiential aspect has become highly popular in recent years for Asian tourists to the extent that utilising virtual tours in order to attract Asian tourists has been investigated (Jokelainen, 2018). Some high latitude areas can retain snow cover until May and June; this, combined with extended periods of 24-hour sunlight, which is generally termed the "midnight sun" further adds to the uniqueness of the Arctic as a tourism destination.

Little knowledge is available on how we design Arctic spaces for Asian tourism. Most knowledge is praxis-based on local tourist destinations, such as Rovaniemi, Finland, for Santa Claus/Christmas resorts. Here we need to understand the motivations of Asian tourists visiting and "consuming" the space of an Arctic city. As most Asian tourists travel to these spaces by foot after being transported to the attraction spaces via other means of transportation such as flights, boats, and/ or buses, it is important to understand the challenges they will face when visiting an Arctic city, especially in the winter season. For tourists, Arctic conditions are unfamiliar and challenging and can result in leisure-time accidents (Lépy et al., 2016). Closely connected trends of busier tourist seasons and concerns about an anticipated shortening of the winter season also lead to a higher tourist concentration at certain periods of the year.

This chapter focuses on how the design of Arctic settlements can enhance Asian mobilities and consumption on foot by walking. To this end, it adopts a framework based on the built environment of Arctic cities, human movement, and climate (See Figure 13.1 adapted from Chapman & Larsson, 2019). The research question for this chapter is: *with particular attention to Asian mobilities consumption, what urban design considerations should be accounted for when planning Arctic cities?* This research question helps address the recognised need for the further study of tourism in Arctic urban settlements (Müller et al., 2020).

The built form of Arctic cities

For Asian tourists in the Arctic, the built environment is the basis of their mobility and consumption. The space of an Arctic city is small; for example, Kiruna, Sweden has a population of around 25,000, and Tromsø, Norway's is approximately 72,000. The public space of settlements can be described in many ways. Aristotle saw space as the container of all things, Newton saw space and time as real things with infinite duration, while Einstein said that the concept of space was a creation of the human imagination (Madanipour, 1996). However, when considering Asian mobilities consumption in the Arctic, Lefebvre (1991) identified three types of space that are important in spatial consideration: conceived space, perceived space, and lived space.

Conceived space is representational space that is a drawn space used to illustrate our built environment (Lefebvre, 1991). Perceived space is how space is understood in psychological terms. For example, whether a space or street is perceived as safe or not. Here Lefebvre's idea was to focus on the perception of everyday space (the daily reality). Lefebvre saw this space as the outcome of

Figure 13.1 Mobilities consumption emerges from the interactions of the urban form, the individual, and the winter season

many processes, both societal and historical (Merrifield, 2014; Pierce & Martin, 2015). Lived space is the experienced space, which cannot be disaggregated from physical space and social circumstance (Fraser, 2008; Madanipour, 2017; Pierce & Martin, 2015).

This is important in understanding Asian mobilities consumption in the Arctic. While the conceived space of settlements (designed space) stays the same all year round, winter changes the form by adding the cover of winter. Accordingly, the urban space becomes a container of the physical dimensions of winter (Chapman, 2018). Together, the combination of the built environment and the climate creates the "perceived space" that the Asian tourist is expecting to experience from tourist information, advertising, etc., that has generated their image of the Arctic. The "lived space", however, is the space the Asian tourist experiences on their visit. This may not reflect their image of an "ideal winter" and can create disappointment. The latter may occur, for example, if there is a lack of snow or when atmospheric conditions impact the likelihood of seeing the Northern Lights.

During city visits, Asian tourists will also seek to understand these settlements by searching for understandable and familiar settlement patterns (Carmona et al., 2010; Kropf, 2014, 2017), that is, the urban morphology and patterns of an Arctic city. Here, "urban" includes all human settlements irrespective of size (i.e. it is not limited to cities), and 'morphology' means all urban forms (Kropf, 2009).

Forms include places, networks of buildings, streets, and spaces that makeup the public realm of a settlement. While these patterns vary in complexity, they all relate to how a spatial system and its movement system are produced for mobilities. Further, when moving by foot, the connectivity that these movement systems produce, that is the degree to which a point or place is connected to other points and places, is a founding principle of good urbanism (Gehl, 2011).

The fact that space in the Arctic is partly constructed by human activity and partly by natural forces is important for the morphology of space in Arctic cities (Chapman, 2021). Here forces that govern aspects of urban form include non-human forces (e.g. natural forces), albeit some aspects of urban form are not consciously shaped, for example, how snow can cover walkways in the city. Here, while the morphological patterns of natural forces create variety in complexity, they relate directly to how a spatial system and its movement system are produced.

At a basic level, tourist maps are likely to illustrate the "built" form of the Arctic city and not the "urban" form that has been reshaped and partly formed by winter. It is the reshaped form the Asian tourist will experience. Such reshaping challenges the basic ideas of urban design, that is, the fixed nature of legibility, townscape, and connectivity. For the Asian tourist, this can make Arctic cities difficult to read, orientate within, and way find.

Human movement: Asian mobilities in the Arctic

How people perceive their environment and move is a crucial aspect of tourism (Lau & McKercher, 2006). Enabling Asian tourists to travel by foot in an Arctic city brings many benefits. Such benefits include general good health and the psychosocial value of enabling the tourist to meet others in Arctic settlements and take part in civic life. Walking also facilitates tourists to fully perceive and experience the local landscape, either the natural or built forms (Fyhri et al., 2009).

Shumway-Cook and Woollacott's (2017) theoretical framework for movement links three interacting factors: the individual (and their capacities, perceptions, and experiences), the task (for example, walking to a destination), and the environment. This framework, which was further advanced in regard to climate and urban form by Chapman & Larsson (2019) (see Figure 13.1) helps unravel the interaction between the environment and people's movement in Arctic cities.

The Asian tourist is a new user of Arctic cities and embarking on a new experience that is exotic and potentially once in a lifetime. This raises problems related to an individual's capacities, perceptions, and experiences. Often, an Asian tourist's perceptions and past experiences from their own home are not relatable to Arctic cities and their harsh environments. Good examples of the individual's capacities, perceptions, and experiences are provided in the chapters by Cheung, Sokolíčková, Kotašková, and Jelínek in this volume.

This lack of familiarity challenges a person's self-efficacy, which is their perceived ability to mobilise their cognitive resources and action in response to situational demands (Bandura, 2017). As their experiences may not be transferable to Arctic conditions, subsequently, the Asian tourist is unlikely to have a collective memory of environmental conditions that they will experience in the Arctic. This situation may impact on their pleasure and desire to take part in tourist activities and also create risk related to injuries (for example, slipping and falling) in the outdoor spaces of Arctic cities. Examples of environmental-conditions collective memory include confidence in auditory sensory clues regarding how crisp the snow sounds under your shoes when walking on packed snow with a good grip, or, in the darkness, when visual clues cannot provide a walker guidance about ice edges when snow piles hinder walkways. Here, in the absence of prior experience, Asian tourists should be given the opportunity to make judgements about local information (seeing how others are succeeding and/or direct encouragement from other persons). This can also be linked to the theory of planned behaviour (Ajzen, 1991). Soft initiatives such as tourist information can facilitate behavioural control or self-efficacy with respect to necessary skills and abilities to perform a specific task (Bandura 2017; Fishbein, 2009). For example, riding a snow-mobile or walking safely on icy walkways with confidence in individual movement allows more attention to be placed on viewing the beautiful night sky and sites, rather than looking down on where to place your feet.

For Asian tourists and consumption in Arctic cities, "soft information" is important in allowing a tourist to appreciate the cultural values and enjoy their stay without either unnecessary fear or risk. Here, also the concept of interpretation can be applied to describe how guided interpretation of the natural and cultural values of a site can provide the tourist with higher satisfaction from a visit to the site. It also supports that age and mobility limitations imply different needs for interpretation (Weng et al., 2020). Aiding tourists' perception of cultural values/Confucian heritage and building self-efficacy in safely moving at and between sites are important for a positive experience.

Winter season in Arctic cities and climate change

The relationship between the state of the earth and human activity has been a subject of discussion throughout history (Tyrrell, 2013). Over time, human life and the environment have developed together and are linked by a causal relationship. The environment is a coupled system that is shaped by life and vice versa (Tyrrell, 2013).This argument is important in the context of this chapter because humans influence the environment and influence climate change.

For the Arctic, climate change is co-produced with human activity, including tourism, and is often described as happening at "double-speed". That is, annual climate warming is being reported at a level almost double that of previous highs in temperature increases (Overland et al., 2019). Climate change in the Arctic is changing the type of winter in Arctic cities that Asian tourists experience (Cai, Ma, & Lee, 2020; Jørgensen & Bertelsen, 2020; Martin & Ren, 2020). This is very significant, as tourists want to experience long predictable winters, and yet they are at risk. Long, cold, and snowy winters are being replaced by warmer winters and more rain.

These factors need to be carefully considered in the design of winter cities for Asian tourism. This situation will also bring new pressures for Asian consumption in the Arctic and will require new strategies to extend the tourist season.

Conditions and contexts for winter city urban design

For Arctic cities, the adapted framework depicted in Figure 13.1 suggests a need to re-balance design activities between "bare-ground" summer conditions and "snow-covered" winter conditions.

Design projects for Arctic cities will need to consider how warmer temperatures with increased levels of precipitation (more of which is likely to fall as rain) will alter mobility choices and perceptions about attractive destinations. In particular, designers may face the challenge of winter conditions that are around 0°C. These conditions are particularly difficult because designers must be aware that precipitation may regularly fall as snow or rain, creating very treacherous surface conditions with a covering layer that may consist of snow, ice, slush, or frozen slush. In this regard the opportunities of shortening the midseason (with rain), strategies can be put forward to extend the benefits of snow. For example, snow can be stored (Lintzén et al., 2019) and used to prolong the "winter season". If used well in the public spaces, preserved snow cover reflecting the sky lights can be retained as part of the winter experience anticipated by tourists.

The main points of focus in urban planning and design are connected routes as well as available and understood public realms for soft mobility (Larsson & Chapman, 2020). These become constantly evaluated in the face of climate change. Overlaying the additional dimension of Asian consumption further highlights the ever-changing, hence, dynamic nature of urban space planning and design. As illustrated in Figure 13.2, the needs of tourism and Asian mobilities consumption should be factored into the design of winter cities. This is in line with the philosophical basis of the Confucian five constants interacting with Veblen's conspicuous consumption (as before, see Chapter 2 of this volume for further discussion). Indeed, when East Asian, primarily inclusive of Chinese, Japanese, and Korean, cultural philosophies are enacted during constant interchanges with dynamic social and physical elements, some elements may need further consideration, and these are presented accordingly in the following section.

Aspects in Arctic winter city urban design in relation to Asian mobilities consumption

Arctic settlements should take a blue-green-white approach to planning that addresses the structure, function, and design of green, blue, and public areas, spaces, streets, and paths when they become white due to snow and ice. Future plans will need to achieve an attractive built environment where transport by walking and biking is prioritised and inviting as an everyday activity, all year round. Plans should focus on winter connections, pathways and soft mobility and formal vehicular infrastructure; public space maintenance and management;

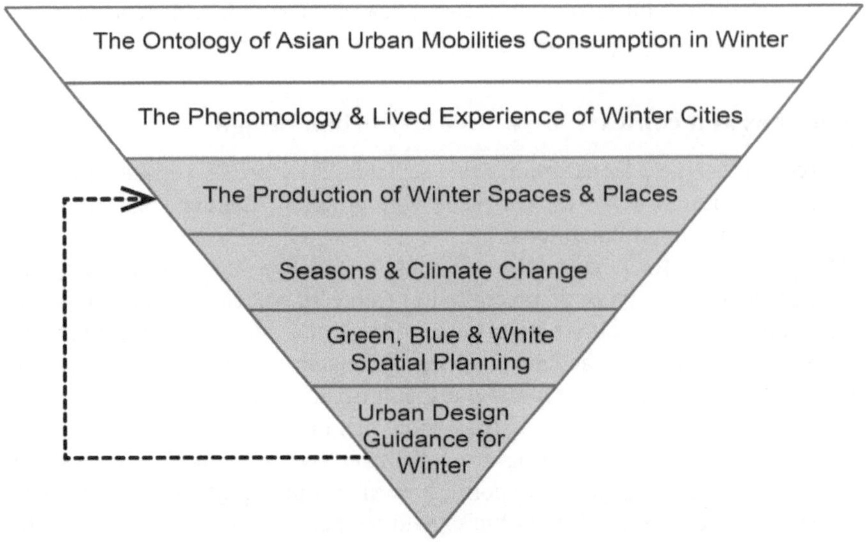

Figure 13.2 Conceptual framework for Asian mobilities consumption and winter city planning and urban design guidance

and the outdoor winter environment. As the winter season is also dark, planners and designers should include the structure, function, and design of lighting. At a technical level, white plans address materials, snow removal, and storage (Chapman et al., 2017). For Asian tourism, this blue-green-white approach to planning should include "soft planning information" on appropriate clothing and footwear for the Arctic, perceived temperature conversions, potential risks, and basic information on how safely to walk on the snow and ice (Larsson, Berggård, Rosander, Gard, 2019).

Specifically, three factors for designing Arctic cities with a specific focus on Asian mobilities consumption can be considered. Areas of possible interactions with Veblen's conspicuous consumption in the context of the Confucian five constants (see Chapter 2 of this volume) are provided with further points of reflection.

Urban form: more compact development patterns can help facilitate Asian tourist mobilities by positioning facilities and destinations closer together. However, urban forms cannot be imported directly from lower latitude areas that do not experience the same seasonal climates as Arctic settlements. Equally, the perceived and lived spaces (Fraser, 2008; Madanipour, 2017; Pierce & Martin 2015) of Arctic cities are dissimilar to those of an Asian urban form with dense populations. It is important to remember that tourist experiences are a product of a finely tuned combination of novelty and the everyday life of things (Franklin, 2003). However, due to harsh environmental conditions unknown to tourists, including those from Asia, planners should place physical safety threats as a priority when designing Arctic cities.

Incorporating a green shift in energy sources into overall tourist experiences by utilising solar energy in designs could also provide Asian tourists with an opportunity to gain insights into Arctic cities' green energy transition. Wisdom (*Chi*), one of the five constants in Confucian values (see Chapter 2), is an entity that is a continuously evolving state in human societies.

Chi may also further prove to be a significant experiential element to Confucian heritage tourists. Since they constitute a quarter of the world's population (Park & Cho, 1995), an understanding of the Confucian heritage of Asian people when they visit Arctic cities may prove to be a valuable asset for sustainable business practice.

Streets and spaces: an area's network of streets can have a strong influence on soft mobility choices. Here, designers should consider how the network of streets and routes could improve the microclimate. Irregular or broken street grid layouts and curved streets can significantly reduce wind and improve the pedestrian environment. Street patterns can support soft mobility by connecting local and more distant destinations together while funnelling and blocking the wind from routes.

Designing slightly wider streets and routes than those found in places that mainly experience bare-ground conditions will make it easier to accommodate soft mobility in winter. Wider streets help ensure enough space is available for both winter cover and pedestrian movement. The section of the streets and public realm can also be designed to move rain, slush, and snow away from areas to enable soft mobility.

The winter conditions tend to disguise the elements of the public realm. One way to overcome bare-ground details being hidden by snow is to raise this information about the public realm, its networks, routes, and designations above the ground level. This can help tourists read the legibility of the townscape and enable easier understanding of the area and its public realm.

The importance of soft information to tourists in the Arctic or a lack of tacit knowledge of the Arctic is further highlighted when tourists encounter language barriers due to illegible signages or sign postings. If communicated properly, signage can guide tourists on appropriate ways (Propriety, *Li*) to consume the public spaces as a starting point of their urban Arctic experiences. Confucian *Li* refers to propriety that leads to certain rituals and rites. Accordingly, Li (2007) suggested that the notion of *Li* should be interpreted as cultural grammar. By observing certain rituals in the way that needs to be performed, *Li* is comparable to another Confucian notion *Ren* (benevolence) that might be interpreted as mastery of a culture. Both should be considered when developing planning strategies for soft information.

Intermediate (indoor-outdoor) spaces at building entrances should also be considered as these provide buffer zones in the winter. These spaces can help enable soft mobility by providing places for people to adjust from being outdoors in ways such as removing winter boots and clothing.

Public realm management: if we want to make a city more attractive for tourists and encourage them not to use motorised transport in the winter, public realm management plans need to be prepared that prioritise the maintenance of the public realm for soft mobility. Design strategies should focus on "designing" the winter soft mobility networks and delivering a connected network of routes that relates to the urban patterns of Asian mobilities consumption and related tourist centres, hubs, and destinations. This could be described as a winter tourism public realm management plan.

The connected network of routes may allow a more sustainable solution not only to tourists, whose physical and experiential capabilities require soft mobilities, but also for Arctic tourism experience providers. A softening of Arctic adventure tourism with Asian tourists in the region has been observed in recent times (see the chapter by Cheung in this volume; Matin & Ren, 2020). Consequently, less-risky or less-weather dependant activities may be designed for Asian tourists, which could demand certain changes in the management of public realm.

Conclusion

Overall, the aim of urban design and city planning in Arctic cities is to create liveable and sustainable cities and places for residents and tourists. However, when considering Asian tourists in the Arctic and their mobilities and consumptions, two key conclusions emerge.

First like all tourists, they are "tourists". This means that they are unlikely to have the individual or collective memory of how to behave in the outdoor environments of an Arctic city. This requires the tourism industry to provide information and soft-planning knowledge of how an Arctic city operates to its tourists. This could include information on clothing, but also "tourism maps" that visualise the winter townscape and layout of the city with its winter cover of snow and ice.

Secondly, city planners and architects should place special focus on "how" the public realm will be used in the winter and how Asian tourists will read the urban environment. This can include designing places with enhanced legibility, adopting new techniques to wayfinding that are not masked by the winter season, or simply, providing information on safe areas, for example, walking on a frozen river.

These soft and hard planning interventions need to be adapted to address climate change and the rapid warming of Arctic cities. Those involved with the long-term planning for sustainable tourism will need to develop strategies that are still attractive to tourists in the long term.

References

Ajzen, I. (1991). The theory of planned behavior. *Organizational Behavior and Human Decision Processes*, *1*(50), 179–211.

Bandura, A. (2017). *Self-efficacy: The exercise of control*. New York: W.H. Freeman and Company.

Cai, Y., Ma, J., & Lee, Y.-S. (2020). How do Chinese travelers experience the Arctic? Insights from a hedonic and eudaimonic perspective. *Scandinavian Journal of Hospitality and Tourism*, *20*(2), 144–165. doi:10.1080/15022250.2020.1744185

Carmona, M., Tiesdell, S., Heath, T., & Oc, T. (2010). *Public Places – Urban Spaces: the Dimensions of Urban Design* (2nd ed.). Oxford: Architectural Press/Elsevier.

Chapman, D. (2021). Is climate a modifier and shape-giver in urban morphology? *Journal of Urban Morphology*, *25*(1), 5–8. doi.org/10.51347/UM25.0007

Chapman, D., & Larsson, A. (2019). Toward an integrated model for soft-mobility. *International Journal of Environmental Research & Public Health*, *16*, 3669. doi:10.3390/ijerph16193669

Chapman, D., Nilsson, K., Rizzo, A., & Larsson, A. (2018). Updating winter: the importance of climate-sensitive urban design for winter settlements. *Arctic Yearbook*, pp. 86–105.

Chapman, D., Nilsson, K., Larsson, A., & Rizzo, A. (2017). Climatic barriers to soft-mobility in winter: Lulea, Sweden as case study. *Sustainable Cities and Society*, *35*, 574–580. doi:10.1016/j.scs.2017.09.003

Fishbein, M. (2009). An integrative model for behavioral prediction and its application to health promotion. *Emerging Theories in Health Promotion Practice and Research*, 215–234.

Franklin, A. (2003). *Tourism: An introduction*. London and Thousand Oaks: Sage.

Fraser, B. (2008). Toward a philosophy of the urban: Henri Lefebvre's uncomfortable application of Bergsonism. *Environment and Planning D: Society and Space*, *26*(2), 338–358.

Fyhri, A., Jacobsen, J., & Tømmervik, H. (2009). Tourists' landscape perceptions and preferences in a Scandinavian coastal region. *Landscape & Urban Planning*, *91*, 202–211.

Gehl, J. (2011). *Life Between Buildings: Using Public Space*. Washington, DC: Island Press.

Jokelainen, N. (2018). Bringing Singaporeans to see Northern Lights: How can virtual reality be utilized in travel marketing? (Master's thesis, Tampera University of Applied Sciences, Tampera, Finland). Retrieved from https://www.theseus.fi/bitstream/handle/10024/146015/Jokelainen_Nelli.pdf?sequence=1&isAllowed=y.

Jørgensen, M. T., & Bertelsen, R. G. (2020). Chinese tourism in the Nordic Arctic – opportunities beyond the economic, *Scandinavian Journal of Hospitality and Tourism*, *20*(2), 166–177. doi:10.1080/15022250.2020.1744186

Jóhannesson, G. T., & Lund, K. A. (2017). Aurora Borealis: Choreographies of darkness and light. *Annals of Tourism Research*, *63*, 183–190.

Kropf, K. (2017). *Handbook of Urban Morphology*. Chichester: John Wiley & Sons.

Kropf, K. (2014). Ambiguity in the definition of built form. *Urban Morphology*, *18*(1), 41–57.

Kropf, K. (2009). Aspects of urban form. *Urban Morphology*, *13*(2), 105–120.

Larsson, A., Berggård, G., Rosander, P. & Gard, G. (2019). Gait speed with anti-slip devices on icy pedestrian crossings relate to perceived fall-risk and balance. *International Journal of Environmental Research and Public Health*, *16*(14), Article ID 2451.

Larsson, A., & Chapman, D. (2020). Perceived impact of meteorological conditions on the use of public space in winter settlements. *International Journal of Biometeorology*, *64*(4), 631–642. doi:10.1007/s00484-019-01852-5

Lau, G. and McKercher, B. (2006). Understanding tourist movement patterns in a destination: A GIS approach. *Tourism & Hospitality Research*, *7*(1), 39–49.

Lefebvre, H. (1991). *The Production of Space*. Oxford: Basil Blackwell.

Li, C. (2007). *Li* as cultural grammar: On the relation between *Li* and *Ren* in Confucius' analects. *Philosophy East & West*, *57*, 311–29.

Lintzén, N., Danvind, J., Melin Söderström, E., Nilsson, K., & Skoglund, P. (2019). Laboratory investigation of different insulating materials used for snow storage. *Journal of Cold Regions Engineering*, *33*(4), 04019012.

Lépy, É., Rantala, S., Huusko, A., Nieminen, P., Hippi, M., & Rautio, A. (2016). Role of winter weather conditions and slipperiness on tourists' accidents in Finland. *International Journal of Environmental Research and Public Health*, *13*(8), 822.

Madanipour, A. (2017). *Cities in Time: Temporary Urbanism and the Future of the City*. London, UK: Bloomsbury Academic, an imprint of Bloomsbury Publishing Plc.

Madanipour, A. (1996). Urban design and dilemmas of space. *Environment and Planning D: Society and Space*, *14*(3), 331–355.

Martin, N. & Ren, C. (2020). Adventurous Arctic encounters? Exploring Chinese adventure tourism. *Scandinavian Journal of Hospitality and Tourism*, *20*(2), 126–143. doi:10.1080/15022250.2020.1743752

Mathisen, T. A., Annema, J. A., & Kroesen, M. (2015). The effects of weather and climate change on cycling in northern Norway. *The European Journal of Transport and Infrastructure Research*. Retrieved from http://doi-org.proxy.lib.ltu.se/10.18757/ejtir.2015.15.2.3072

Merrifield, A. (2014). *The New Urban Question*. London: Pluto Press.

Morata, B., Cavalieri, C., Rizzo, A., & Luciani, A. (2020). Territories of extraction: Mapping palimpsests of appropriation. *Urban Planning*, *5*(2), 132–151. doi:10.17645/up.v5i2.2901

Müller, D. K., Carson, D. A., de la Barre, S., & Granås, B. (2020). *Arctic Tourism in Times of Change: Dimensions of Urban Tourism*. Copenhagen: Nordic Council of Ministers.

Overland, J., Dunlea, E., Box, J. E., Corell, R., Forsius, M., Kattsov, V., … Wang, M. (2019). The urgency of Arctic change. *Polar Science*, *21*, 6–13. doi:10.1016/j.polar.2018.11.008

Parai A., & Dutt A. K. (1994). Perspectives on Asian urbanization: An East-West comparison. In Dutt A. K., Costa F. J., Aggarwal S., & Noble A. G. (Eds.), *The Asian City: Processes of Development, Characteristics and Planning*. The GeoJournal Library (vol 30). Dordrecht: Springer. doi.org/10.1007/978-94-011-1002-0_24

Park, I., & Cho, L. (1995). Confucianism and the Korean family [Special issue]. *Journal of Comparative Family Studies*, *26*(1), 117–134.

Pellegrini, A. (2011). *A new approach to urbanization in Asia 2050: Realizing the Asian century*. In Kohli, H. S., Sharma, A., & Sood, A. (Eds.), *Realizing the Asian Century* (129–150). New Delhi: SAGE Publications India. doi:10.4135/9781446270349.n10

Pierce, J., & Martin, D. (2015). Placing Lefebvre. *Antipode*, *47*(5), 1279–1299.

Pressman, N. (2004). *Shaping Cities for Winter, Climate Comfort and Sustainable Design*. British Columbia: Winter Cities Association.

Shumway-Cook, A., & Woollacott, M. H. (2017). *Motor Control: Translating Research into Clinical Practice* (5th ed.). Philadelphia, PA: Wolters Kluwer.

Stokkan, K. A., & Reiter, R. J. (1994). Melatonin rhythms in Arctic urban residents. *Journal of Pineal Research*, *16*, 33–36.

Tyrrell, T. (2013). *On Gaia: A critical Investigation of the Relationship Between Life and Earth*. Princeton, Oxford: Princeton University Press.

Varentsov, M., Konstantinov, P., Baklanov, A., Esau, I., Miles, V., & Davy R. (2018). Anthropogenic and natural drivers of a strong winter urban heat island in a typical Arctic city. *Atmospheric Chemistry and Physics 18*(23): 17573–17587. doi:10.5194/acp-18-17573-2018

Weng, L., Liang, Z., & Bao, J. (2020). The effect of tour interpretation on perceived heritage values: A comparison of tourists with and without tour guiding interpretation at a heritage destination. *Journal of Destination Marketing & Management, 16*, 100431.

Zhang, Q., & Deng, X. (2017). *Urban Development in Asia: Pathways, Opportunities and Challenges*. Singapore: Springer.

Part 5

Closing

14 Sustainable mobilities consumption and Asia in a global Arctic

Observations and ways forward

Young-Sook Lee

Sustainable Asian mobilities consumption in a global Arctic: An interplay between science and politics

Until recently, understanding the physically changing Arctic has been mostly led by natural scientists. The need for a cross-cutting (AMMAP, 2017), cross-sectoral, and cross-scale understanding of the changes in Arctic environments has been increasingly acknowledged. However, while the necessity for integrated disciplinary work and understanding is acknowledged, the tendency to prioritise natural scientific research regarding the changing Arctic continues to dominate and be practised. This is evident in the recent Arctic research expedition by Mosaic. As the largest exploration expedition to the Arctic, more than 300 natural scientists participated from 20 countries, gathering and evaluating important measurement data from late 2019 through to the end of 2020. Consequently, the expedition time included the Arctic winter season. Participants anticipated that the gathered fundamental natural scientific knowledge would contribute towards a better understanding of climate change in the Arctic regions and beyond. The mission was led by the Alfred Wegener Institute, Helmholtz Center [*sic*] for Polar and Marine Research (AWI). Drawing on different media sites and social media platforms, this expedition was followed by many people in various countries. Such a following sparked a great deal of interest in everyday life by non-Arctic experts regarding Arctic natural scientific endeavours. Nevertheless, for the various Arctic fora involved, synchronising research outputs, and reporting results simultaneously and in relevant spaces in order to make changes in everyday life were and are acknowledged as the biggest challenges to overcome (Mosaic, 2019; Arctic Frontiers, 2021). In this book, the importance of synchronising research results was also echoed in Chapter 11 by Sokolíčková. In presenting the transformation of Svalbard's industry base from mining to tourism, the chapter argued the essential need to integrate extant knowledge that is shaping and shaped by the natural and social sciences arenas. Additionally, Chapter 13 by Chapman, Lee, and Larsson provided architectural input into urbanising the Arctic and its spatial consumption by visiting Asian tourists. Chapman et al. argued that it is essential to integrate knowledge and practice when it comes to designing urban forms as

DOI: 10.4324/9781003039518-14

well as addressing winter season challenges and dimensions of Asian tourists' mobilities.

While the integration of natural and social sciences research has been a challenge, researchers in mobilities studies have made attempts to link the impacts on physical environments and mobilities. For example, Runge, Daigle, and Hausner (2020) traced the footprint of Arctic tourism. Extracting geotagged locations from over 800,000 photos on Flickr, the study identified that the footprint of summer tourism quadrupled during 2006 and 2016 and that winter tourism increased over 600% during the same period. Another study researched geolocated travel blogs and identified some spatio-temporal traits of Chinese tourists in Nordic countries (Zheng, Mou, Zhang, Makkonen, & Yang, 2021). The spatio-temporal traits identified by a network analysis of Chinese tourists visiting the Arctic indicated that Iceland demonstrated a larger and stronger trait than other Nordic countries. This strong spatio-temporal tendency of Chinese tourists clearly resonates in Chapter 12 written by Vilhjálmsdóttir, Gunnarsdóttir, and Black. In their research, Vilhjálmsdóttir et al. identified and studied the imbalance between the rate of growth of Chinese tourists and the development and provision of adequate experiential products for newly arriving Chinese visitors. Beyond the Arctic level, Hopkins & Higham (2016) convincingly advocated that the pathway to a low carbon mobility transition could only be paved by multi-sectoral, multidisciplinary, and multi-scale efforts. Chapter 6 by Eikeland and Chiesa in this volume specifically contributes towards linking efforts between Arctic environments and the cruise industry's carbon footprint. Taking the city of Tromsø in Northern Norway as a case, the chapter proposed a green energy source for cruise ships that would bring their carbon footprint to zero. Elsewhere, in Chapter 7, by drawing on a case of the Northeast passage that connects China and the Arctic, Ojala, Räikkönen, and Kiiski discuss the potential of the new cruise passage. By tracing traffic statistics and tourism experiences offered by Arctic experiences operators, the chapter investigated the viability of cruise experience products for the Chinese tourist market.

Not unexpectedly, climate change has drawn the world's attention to the Arctic and its governing body, the Arctic Council. The latter has been constantly adapting to the changing magnitude and levels of its policy division and implementation. Relatedly, the Global-Arctic Mission Council was formally launched in Shanghai at the 2019 Arctic Circle China Forum (Heininen, 2021). The establishment of the global-level Arctic Mission Council demonstrates the expanding political influences and impacts of Arctic matters in regions beyond the Arctic. The political expansion of the Arctic is noticeable and witnessed by the augmented state of Arctic tourism politics. When Hall's (1994) book on international tourism politics was published, the political aspects of tourism were closely aligned with "doing business" *per se*. Earlier, Richter's (1989) volume about the politics of tourism in Asia focused on facilitating tourism product developments and business transactions by articulating both the good and bad sides of each. Chapters 3, 4, and 5 in this volume investigated political aspects of Asian mobilities consumption in the Arctic. These chapters present a significantly altered dimension of politics from that of the earlier works. Specifically, Chapter 3 by Björner, Dybtsyna, and Lundberg sought to identify shared ground in Arctic tourism development strategies between three Arctic nations (Sweden, Norway, Finland) and those of China. The research focus itself is testimony to the heightened aspect of political dimensions in Arctic tourism. The fact that the tourism sector is placed and viewed in Arctic political strategies in Sweden, Norway, Finland, and China goes beyond the business sector realm. Moreover, the fact demonstrates that Arctic tourism is not the same entity that was experienced earlier in other parts of the globe. A different example focusing on the Far East and the Russian Arctic is provided in Chapter 4 by Kobzeva. She examined the impact of a Sino-Russian political partnership on epistemic community tourism development. Kobzeva found that Chinese tourism to Russia plays a role in Sino-Russian relationships, albeit currently there are challenges for Russian tourism providers in meeting Chinese tourists' meaningful consumption. Further evidence of the overarching force of political dimensions within Arctic tourism is presented in Chapter 5 by Reigner, Hovgaard, and Gaini. These authors focused on Greenland and the Faroe Islands and their respective plans for sustainable Arctic tourism bearing in mind Chinese inbound tourist numbers. Indeed, Reigner et al. convincingly argue that engaging with Chinese tourists in Greenland and the Faroe Islands through a sustainable Arctic tourism is not located in the mere "tool to do business" level but in a China-Russia political partnership. This focus in Chapters 3, 4, and 5 on the political aspects of Arctic tourism demonstrates that evidence from industry and government documents highlights the heightened importance of politics in Arctic tourism compared to international tourism beyond the Arctic. Notably, the book's attention to the changing political dynamics in the Arctic is concentrated on China. This focus reflects China's political power expansion through the agent of Arctic tourism into the Arctic where China does not claim territorial sovereignty (Bennett & Iaquinto, 2021). Recognition of such power expansion by political scientists further affirms that Arctic tourism is a political tool used in multiple ways.

Framing Asian mobilities consumption according to Veblen: Asian conspicuous consumption in the Arctic

At the beginning of this book, in order to understand Asian mobilities consumption in the Arctic, a specific focus was Veblen's conspicuous consumption. This focus was intentional in design and was derived from the seemingly agreed and accepted understanding that Asian consumers can aptly be described and understood as Veblen's conspicuous consumers (see Li et al., 2019; Park & Reisinger, 2009; Phau & Teah, 2009; Zakaria, Wan-Ismail & Abdul-Talib, 2020). Another focal point has been in approaching an "Asian" perspective beyond solely a Chinese one to include East Asian peoples with Confucian cultural philosophical heritage. Subsequently, the Confucian five constants were chosen to provide a further concrete focal point upon which analyses could be based. This fitted into the boundaries of the current volume, which essays to understand Asian mobilities consumption in Arctic destinations.

Based on the chapters in this book, what is apparent is that Arctic mobilities are an agent of changing Arctic politics, which encompasses Arctic peoples and environments, multi-disciplinary fields, multi-sectorial engagements, and multi-layered policy divisions and implementations. The dynamic nature of Arctic politics is becoming increasingly globalised in the 21st century and this globalisation has drawn the attention of international powers and their subsequent interests in the Arctic. Further, drawing on the chapters in this book, I would argue that the labelling of Asian mobilities consumption as seemingly an un-questionably accepted form of Veblen's conspicuous consumption leaves room for erroneous understanding. Indeed, when Beng-Haut (2000) edited a volume titled as *Consumption in Asia*, one of the book's focal arguments was that consumption in East and Southeast Asia is very much tied to each country's respective political economies. About 20 years after that claim, Asian mobilities consumption in the Arctic appears to hold this trait. Not only is the trait held, but the trait appears to have further amplified due to the tourism sector in the Arctic being significantly more aligned with political dimensions than in other destinations on the globe as has been already mentioned. Another Asian consumption trait provided was "traditional" thrift against "excess" in consumption (Beng-Haut, 2000). The second trait explains the conflicting situation where traditional values in Asia purport thrift while the contemporary lifestyle taken up by the new rich in Asia seeks excessive consumption, going against traditional cultural values. Supporting the "traditional" thrift against "excess" trait, other research by Sigurðsson (2014) argued that Confucian values are situated on the opposite side of the spectrum to excessive consumption. When Sigurðsson (2014) conducted his study specifically investigating the Chinese philosophical perspectives on consumerism, the Chinese traditional philosophy represented by Confucianism was found to discourage rather than stimulate non-essential consumption. The traditional thrift versus excessive consumption thesis is on the opposite side of some studies' findings, which claimed that traditional cultural philosophy in Asia, e.g. Confucianism, had provided the cultural base for Asian consumers' excessive and conspicuous consumption (see Podoshen, Lu, and Zhang, 2011; Wong & Ahuvia, 1998; Park, Rabolt, & Jeon, 2008).

Therein, we are presented with two opposing research findings on the traditional East Asian cultural philosophy of Confucianism and its influence on today's East Asian conspicuous consumption. How should this disagreement in earlier studies be interpreted and made meaningful in this book that searches for an understanding of Asian mobilities consumption in the Arctic?

It may benefit us to view the East Asian cultural philosophy as part of dynamic and shared thought systems rather than a rigid and exclusive one in relation to other cultural philosophies of the world. That is to say, when we view the Confucian five constants, i.e., *Li* or propriety, *Ren* or benevolence, *Chi* or wisdom, *Xin* or sincerity, and *Yi* or righteousness, these cultural philosophical values are not exclusive to Confucian heritage East Asians. Instead, we see these cultural philosophical values as universal, although they might be more structured and discernible in Confucian heritage East Asians. In this vein, we can postulate that the dynamic nature of the Confucian cultural philosophy changes accordingly when it interacts with Western values. Such interaction and change influence the ways in which the fundamental tenet of Confucianism is practised in today's East Asians' consumption. In particular, Confucian influences on today's consumption can be viewed in conjunction with certain adopted Western values and practices. Specifically, in the realm of mobilities consumption, individual-level tastes and preferences, expressed through symbolic choices in the West, may arguably have been overtaken by Confucian fundamental values for East Asian consumers. For example, expressing oneself through mobilities consumption by Western-based individual choice or taste for travel destinations or activities might better be interpreted as adoption of the dynamically changing Confucian concept of *Li* or propriety, wherein individual choice has been replaced by the society-chosen Confucian concept of *Li* or propriety. Subsequently, what is largely a matter of personal choice and preference in the West might instead be a matter of a society-rooted and approved method of expressing oneself in today's Confucian heritage society. Thus, it may be argued that symbolic representations of mobilities consumption by East Asians today continue to denote deeply rooted elements of Confucian values, as in the case of *Li* or propriety being accepted conduct instead of individual expressions of taste and preferences.

An Arctic standpoint

With regard to the chapters presented in this book, there are three points to which I wish to draw the reader's attention. First, this book attempts to understand Asian mobilities consumption in the Arctic from an Arctic point of view. The starting point of the Arctic and Arctic tourism and mobilities was clearly an Arctic emic standpoint. With this standpoint, the authors in this book saw the changing political discourses and attendant mobilities consumption patterns that were observable in Arctic destinations through the eyes of Arctic guides, local residents, politicians, tourism experiences providers, engineers, and architects of Arctic destinations. The chapter by Kotašková and Jelínek, which contemplates the notion of "Asianness" from Arctic guide experiences in Svalbard, clearly posits that the standpoint of understanding Asian mobilities consumption is grounded from the Arctic side. The two authors reported that in Arctic adventure guiding experiences in Svalbard, endeavours to approach Asianness from an ethnicity dimension proved insufficient and were replaced by evaluations of tourists' performances as a better indicator to understand these new tourists to and in the Arctic.

Another intriguing aspect of an Arctic-imposed version of Asian mobilities consumption relates to the northern lights. Notably, Asian tourists hold an ardent zeal to search for the lights when visiting Arctic destinations. As a consequence of such search zealousness, Arctic experience providers recognise that the northern light searching activities are a good experience product. Subsequently, they prepare and accommodate northern lights search experiences. However, the ways in which the Arctic experience providers believe the meanings and motivations for the search for the northern lights by visiting Asian tourists is a sheer Arctic interpretation of the phenomenon. Indeed, it is widely understood by Arctic destination experience providers that the motivations for Asian tourists' searches for the northern lights is due to a Japanese cultural belief. A belief that says when a couple conceives a baby under the northern lights, the baby is believed to grow up to become smart and healthy. However, this belief is an unfounded claim. The origin of the concocted belief is a television series, *Northern Exposure*. The CBS television series ran from the summer of 1990 to its series finale in July 1995, a total of 110 episodes (Plasketes, 2015). In one of the episodes, the supposed Japanese folklore was created and scripted by Jeff Vlaming, the screenwriter for the 1992 episode in which the myth was aired (Garrett, 2007). This American-created and Arctic-adopted belief survives to date whereby northern lights tourism is offered to visiting Asians in other Arctic destinations. The myth/belief is also demonstrated in preparations made by the Russian government for the arrival of Chinese tourists to the Russian Arctic. As related by Kobzeva in Chapter 4, one of the Russian government reports with a focus on mass consumption of nature and history was specifically entitled as "Conceive under the northern lights. Due to Eastern beliefs, the number of Chinese tourists in Murmansk has grown 100 times" (Pertsev, 2019). This report is referred to in order to support "Aurora hunting and glass-wall hotels" as part of the government's development plan for Asian tourists. The report serves to perpetuate the myth through its title, strategies, and linkage to the growing number of Chinese tourists in Murmansk. Subsequently, the report demonstrates that some Arctic destination development has adopted the myth as if it is a proven knowledge. Additionally, in the report, the supposedly Japanese cultural belief was extrapolated to assume that Chinese tourists shared the same belief. These identified examples of Arctic-standpoint observations are a testimony to the nature of the early stages of interactions between Asia and the Arctic as enabled by Arctic tourism.

Li or propriety and conspicuous mobilities consumption in the Arctic

Given the early stages of Asian mobilities consumption in and of Arctic destinations, there appears to be certain interactive phenomena that may be claimed as linking Confucian *Li* or rites and rituals of propriety and Veblen's notion of conspicuous consumption. As noted in Chapter 2, *Li* or rites, rituals of propriety, is one of the central concepts in Confucianism. Because of its centrality and interconnected nature with other Confucian concepts, particularly with *Ren* or benevolence and human excellence, some inconsistent interpretations have been observed. However, understanding *Li* or propriety as a cultural grammar and *Ren* as the mastery of a culture (Li, 2007) provides us with a clearer conceptualisation. It can be further explained that *Li* or propriety serves as the means for us to cultivate and express *Ren* or benevolence. Therefore, *Li* or propriety "can change over time and some of the changes can be justified" (Li, 2007, p. 312).

This Confucian *Li* or propriety notion is not entirely alien to Veblen's 19th-century America. In fact, in various parts of his writing, Veblen talks about how a society expects a gentleman [*sic*] to eat, drink, dress, behave, and notably waste his time and resources in a socially correct way. Over time, this correct way becomes habitus for a certain class of the people of a society, the class with a higher culture than the rest of a society. As such, the correct way of expressing oneself brings acceptance, recognition, and prestige to those who express them, e.g. appreciation of particular alcoholic beverages, high commands of obscure languages, and an extensive knowledge on an extremely narrow field with little practical benefits to sustain one's everyday life. Similar to *Li* or propriety being able to change over time (Li, 2007), Veblen's correct way of wasting time and resources has been recognised as changing over time. This recognition is clearly represented in the two types of conspicuous consumptions detailed in Chapter 2 of this book. Reflecting the traits of *Li* or propriety along with Veblen's conspicuous consumption idea, additional two chapters in this book provided examples of Asian mobilities consumption behaviours at this stage of their consuming Arctic destinations. Mathisen and Søreng in Chapter 9 recounted that visiting and consuming the Thai King's legacy in North Cape as inscribed upon The Stone was linked to status. Hence, there is a certain prestige assigned to contemporary Thai visitors to the site. It is worthy of being reminded that while Thailand is denoted as a largely Buddhist nation, Confucianism is practised throughout Thailand (Niffenegger, Kulviwat & Engchanil, 2006). Cheung in Chapter 10 also provided his own lived experiences as a guide and explained how keen the Chinese adventure tourists were in Arctic destinations with regard to seeking status validation from their Arctic adventure experiences.

Chi or wisdom and conspicuous mobilities consumption in the Arctic

The Confucian *Chi* or wisdom is related to the need for constant self-cultivation (Tu, 1985; Tucker, 1998) in order to reach and maintain the state of *Yi* or righteousness through the practice of *Li* or propriety. In this sense, the Confucian *Chi* or wisdom may be further linked to "learning for worldly development" (Lee & Prebensen, 2018, p. 222). As Confucian values interact with changing cultures over time as well as adapt in time- and space-specific contexts, one witnesses an East Asian specific emphasis on education for worldly developments in the 21st century (Choi & Nieminen, 2013). This yearning for learning for worldly development, placed in the context of a Western standpoint, may bring out some complex issues, such as intellectual property and copyright law. Indeed, Confucianism has been condemned for blocking or slowing down intellectual property laws at an international level for decades (Yu, 2014). The concept of intellectual property did not exist in the time of Confucius. A recent change in the landscape of intellectual property has occurred in East Asia, with Japan, China, and South Korea being among the top five countries to file international applications through the Patent Cooperation Treaty. Subsequently, any condemning link between Confucian values and prevention of intellectual property laws at the international level needs more nuanced research (Yu, 2014).

This complex terrain of intellectual property and copyright matters was reflected, albeit briefly, in Chapter 11 by Sokolíčková. Recounting the business aspects of hosting Chinese tourists in Svalbard, the interviewee mentioned "copy culture" in China and Chinese businesses' practice of "copy with pride". In the overall discussion of doing business with China with respect to hosting Chinese tourists in Svalbard, the captured notion was relatively small in magnitude. Nonetheless, it is an element that is present and corroborated in different areas of disciplinary efforts. Thus, in the early stages of Asian mobilities consumption in and of the Arctic, it is significant to be aware of this phenomenon of "copy culture", as mentioned by the informant in Chapter 11. While the *Chi* or wisdom concept in my interpretation is most closely related to the copying and learning aspect, the concepts of *Xin* or sincerity and *Yi* or righteousness may also have relevant elements when it comes to copying and learning issues. These potential links should be further researched in efforts to better understand Asian mobilities consumption. Figure 14.1 presents the observed phenomena as the result of the interaction between the Confucian five constants and Veblen's conspicuous consumption in Arctic destinations.

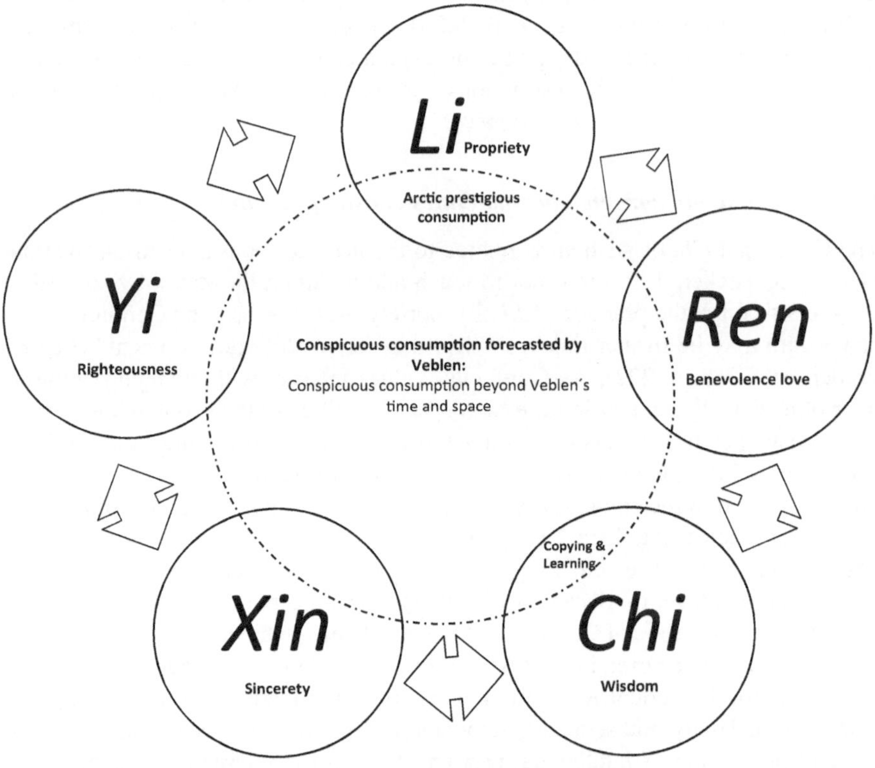

Figure 14.1 An Arctic standpoint on Asian mobilities consumption: when the Confucian five constants meet Veblen's 21st-century conspicuous consumption in Arctic destinations

Post-Covid era mobilities and the future of Asian mobilities consumption in the Arctic

The Covid-19 pandemic has put a near-complete stop to global tourism as well as to Asian mobilities consumption in the Arctic. This book was initially planned prior to the Covid-19 pandemic. At that time, Asian mobilities consumption in and of Arctic destinations was envisaged to progress according to changing Arctic politics. With the unforeseen pandemic affecting the entire globe and with the end of it still yet to come, previously envisaged Asian mobilities consumption is best understood as exhibiting an altered form. Altered forms of global tourism and mobilities have been discussed in various arenas with respect to challenges to the continued growth of volume (Gössling, Scott & Hall, 2021), the importance of local context, and the development of complex globalisations along with getting used to new normal everyday life, travel, and mobilities (Brouder, Teoh, Salazar, Mostafanezhad, Pung, Lapointe, Higgins Desbiolles, Haywood, Hall, & Clausen, 2020). While trying to forecast the exact pathway of Asian mobilities consumption in the Arctic might not be fruitful, it is reasonable to foresee that Arctic tourism as an epitome of Arctic politics may not change in a significant way. The changing politics over the governance and access to the fast-melting Arctic and its resources might be affected through the experience of the pandemic. However, the interplays between scientific knowledge and Arctic politics indeed are expected to be witnessed in the dynamics of Asian mobilities consumption in a changing Arctic.

References

Arctic Monitoring and assessment Programme, AMAP. (2017). *Snow, Water, Ice and Permafrost in the Arctic (SWIPA) 2017*. Oslo, Norway: Arctic Monitoring and Assessment Programme (AMAP).

Arctic Frontiers (2021). *Arctic Frontiers 2021*. Retrieved from https://www.arcticfrontiers.com/arctic-frontiers-2021/

Beng-Huat, C. (2000). Consuming Asians: Ideas and Asians. In C. Beng-Huat (Ed.), *Consumption in Asia Lifestyles and Identities* (pp. 1–34). New York: Routledge.

Bennett, M. M., & Iaquinto, B. L. (2021): The geopolitics of China's Arctic tourism resources. *Territory, Politics, Governance*. doi:10.1080/21622671.2021.1887755

Brouder, P., Teoh, S., Salazar, N. B., Mostafanezhad, M., Pung, J. M., Lapointe, D., & Clausen, H. B. (2020). Reflections and discussions: Tourism matters in the new normal post COVID-19. *Tourism Geographies*, *22*(3), 735–746. doi:10.1080/14616688.2020.1770325

Choi, S. H-J., & Nieminen, T. A. (2013). Factors influencing the higher education of international students from Confucian East Asia. *Higher Education Research & Development*, *32*(2), 161–173.

: Routledge.

Garrett, J. (2007, 2nd of March). The cold show in Fairbanks, Alaska. *The Future of Travel*. Retrieved from: https://www.nytimes.com/2007/03/02/travel/escapes/02Alaska.html.

Gössling, S., Scott, D., & Hall, C. M. (2021). Pandemics, tourism and global change: A rapid assessment of COVID-19. *Journal of Sustainable Tourism*, *29*(1), 1–20. doi:10.1080/09669582.2020.1758708

Hall, C. M. (1994). *Tourism and Politics: Policy, Power and Place*. Chichester: Wiley.

Heininen, L. (2021). The Global-Arctic mission council. *Arctic Circle Journal*. Retrieved from http://www.arcticcircle.org/Media/Arctic-Circle-Journal03LHeininen-GlobalArctic.pdf

Hopkins, D., & Higham, J. (2016). Transitioning to low carbon mobility. In D. Hopkins & J. Higham (Eds.), *Low Carbon Mobility Transitions* (pp. 1). Oxford: Goodfellow Publishers Ltd.

Lee, Y.-S., & Prebensen, N. K. (2018). Value creation and co-creation in tourist experiences: An East Asian cultural knowledge framework approach. In N. K. Prebensen, J. S. Chen & S. U. Muzaff (Eds.), *Creating experience value in tourism* (pp. 215–227). Selangor, Malaysia: CAB International.

Li, C. (2007). *Li* as cultural grammar: On the relation between *Li* and *Ren* in Confucius' *Analects*. *Philosophy East & West*, *57*, 311–29.

Li, J., Guo, S., Zhang, J.Z., & Sun, L. (2019). When others show off my brand: Self-brand association and conspicuous consumption. *Asia Pacific Journal of Marketing and Logistics*, *32*(6),1214–1225. Retrieved from https://doi.org/10.1108/APJML-04-2019-0225

Mosaic (2019). About Mosaic. Retrieved from https://follow.mosaic-expedition.org/site/about/.

Niffenegger, Phillip, Kulviwat, Songpol, & Engchanil, Napatsawan. (2006). Conflicting cultural imperatives in modern Thailand: Global perspectives. *Asia Pacific Business Review*, *12*(4), 403–420. doi:10.1080/13602380600571211

Park, H-J.; Rabolt, N.J., & Jeon, K.S. (2008). Purchasing global luxury brands among young Korean consumers. *Journal of Fashion Marketing and Management*, *12*(2), 244–259.

Park, K-S., & Reisinger, Y. (2009). Cultural differences in shopping for luxury goods: Western, Asian, and Hispanic tourists. *Journal of Travel & Tourism Marketing*, *26*(8), 762–777. doi:10.1080/10548400903356152

Pertsev, A. (2019, December 30). Conceive under the northern lights. Due to Eastern beliefs, the number of Chinese tourists in Murmansk has grown 100 times. *Meduza*. Retrieved from https://meduza.io/feature/2019/12/30/zachat-pod-severnym-siyaniem (In Russian).

Phau, I., & Teah, M. (2009). Devil wears (counterfeit) Prada: A study of antecedents and outcomes of attitudes towards counterfeits of luxury brands. *Journal of Consumer Marketing*, *26*(1),15–27. doi: https://doi.org/10.1108/07363760910927019

Plasketes, G. (2015) The Cicely, Alaska Soundtrack: Northern Exposure's Enduring Musical Mark on Television. *Journal of Popular Film and Television*, *43*(2), 54–62. doi:10.1080/01956051.2014.997659.

Podoshen, J. S., Li, L. & Zhang, J. (2011). Materialism and conspicuous consumption in China: A cross-cultural examination. *International Journal of Consumer Studies*, *35*(1), 17–25.

Richter, L. (1989). *The Politics of Tourism in Asia*. Honolulu: University of Hawaii Press.

Runge, C.A. Daigle R.M., & Hausner, V.H. (2020). Quantifying tourism booms and the increasing footprint in the Arctic with social media data. *PLoS ONE 15*(1): e0227189. doi:10.1371/journal.pone.0227189.

Sigurðsson, G. (2014). Frugalists, anti-consumers, and prosumers: Chinese philosophical perspectives on consumerism. In A. Hulme (Ed.), *The Changing Landscape of China's Consumerism* (pp. 125–149). Oxford: Elsevier.

Tu, W-M. (1985). *Confucian Thought: Selfhood as Creative Transformation.* Albany: State University of New York Press.

Tucker, M. E. (1998). Religious dimensions of Confucianism: Cosmology and cultivation. *Philosophy East and West*, *48*(1), 5–45

Wong, N. Y., & Ahuvia, A. C. (1998). Personal taste and family face: Luxury consumption in Confucian and Western Societies. *Psychology and Marketing*, *15*(August), 423–441.

Yu, P. K. (2014). Intellectual property and Confucianism. In I. Calboli & S. Ragavan (Eds.), *Diversity in Intellectual Property: Identities, Interests and Intersections* (pp. 1–19). Cambridge University Press. Retrieved from https://ssrn.com/abstract=2506384

Zakaria, N., Wan-Ismail, W-N.A., & Abdul-Talib, A-N. (2020). Seriously, conspicuous consumption? The impact of culture, materialism and religiosity on Malaysian Generation Y consumers' purchasing of foreign brands. *Asia Pacific Journal of Marketing and Logistics*, Ahead-of-print. Retrieved from https://doi.org/10.1108/APJML-07-2018-0283

Zheng, Y., Mou, N., Zhang, L., Makkonen, T. & Yang, T. (2021). Chinese tourists in Nordic countries: An analysis of spatio-temporal behavior using geo-located travel blog data. *Computer, Environment and Urban Systems*, *85*, (1–13).

Index